Essential Electronics

Introduction to Solid State Devices

Lem Ibbotson

Former Principal Lecturer at University of East London
Former Open...

A member of the Hodder Headline Group
LONDON • SYDNEY • AUCKLAND

First published in Great Britain in 1997 by
Arnold, a member of the Hodder Headline Group,
338 Euston Road, London NW1 3BH

Whilst the advice and information in this book is believed to be true and
accurate at the date of going to press, neither the author[s] nor the publisher
can accept any legal responsibility or liability for any errors or omissions
that may be made.

British Library Cataloguing in Publication Data
A catalogue record for this book is available from the British Library

Library of Congress Cataloging-in-Publication Data
A catalog record for this book is available from the Library of Congress

ISBN 0 340 66275 1

Typeset in 10½ pt Times by Wearset, Boldon, Tyne and Wear.
Printed and bound in Great Britain by J. W. Arrowsmith Ltd., Bristol

Contents

Series preface

In recent years there have been many changes in the structure of undergraduate courses in engineering and the process is continuing. With the advent of modularization, semesterization and the move towards student-centred learning as class contact time is reduced, students and teachers alike are having to adjust to new methods of learning and teaching.

Essential Electronics is a series of textbooks intended for use by students on degree and diploma level courses in electrical and electronic engineering and related courses such as manufacturing, mechanical, civil and general engineering. Each text is complete in itself and is complementary to other books in the series.

A feature of these books is the acknowledgement of the new culture outlined above and of the fact that students entering higher education are now, through no fault of their own, less well equipped in mathematics and physics than students of ten or even five years ago. With numerous worked examples throughout, and further problems with answers at the end of each chapter, the texts are ideal for directed and independent learning.

The early books in the series cover topics normally found in the first and second year curricula and assume virtually no previous knowledge, with mathematics being kept to a minimum. Later ones are intended for study at final year level.

The authors are all highly qualified chartered engineers with wide experience in higher education and in industry.

R G Powell
Jan 1995
Nottingham Trent University

Preface

The theory of how semiconductor devices work is inevitably complicated, depending as it does on an understanding of electrical conduction processes in solids. Within this constraint I have tried to make the account as easy to understand as possible by concentrating on important functions, giving second-order complications marginal consideration only. The approach that I have chosen to use is based on electrostatics – that is, the relationship of electric fields in the solid to the distribution of charges.

The demand on you for a reasonable fluency in algebra could not be avoided, and I have assumed that you are familiar enough with electronic circuits to understand the main uses to which the devices are put.

The choice of content and style of development come from my experience in teaching this subject for the Open University and subsequently as a visiting lecturer at Leicester University. Most of us owe an intellectual debt to others, and in this case I should like to acknowledge mine to my colleagues at the Open University, in particular John Sparkes and David Gorham, and to John Fothergill of Leicester University.

Lem Ibbotson
April 1996

1 Introduction

The purpose of this book is to explain how some semiconductor devices and circuits work and why they are designed the way they are. I shall assume that you are familiar with circuit theory, and I hope that you know something about electric field theory, although I shall give explanations as I go along.

The book starts with a short chapter (Chapter 2) on conduction in metals, mainly copper. There are two reasons for this: first you are almost certainly familiar with the behaviour of copper conductors, and second semiconductor integrated circuits are still likely to include metal sections (although possibly aluminium). After having established some of the inner processes of conduction in metals I then go on to examine the slightly more complicated effects in silicon.

The range of materials and devices covered is limited, but I hope that the general principles established will allow you to understand descriptions of other devices and systems, and more detailed or more advanced accounts, afterwards.

The way that devices are made is at least as interesting and important as the way they work: in most cases devices became available when the sophisticated and difficult processes required to make them were perfected. I do not, however, intend to give more than passing reference to these processes; they are covered in a very readable manner in, for example, *An Introduction to Semiconductor Microtechnology*, second edition, by D.V. Morgan and K. Board, published by John Wiley and Sons.

Every now and then in the text I ask a question or suggest that you carry out a calculation. The answer occurs immediately afterwards, so you can read straight through if you like, but if you do what is asked before reading the answer it will help you to learn. There are also exercises at the ends of chapters, with answers at the back of the book. I strongly recommend you to do these.

The summaries at the ends of chapters are more than simple reiteration; they also draw conclusions and emphasize the relative importance of ideas. I hope you will find them helpful in revision if you are preparing for an examination.

1.1 SMALL AND LARGE NUMBERS

In the text I shall be dealing with very small and very large numbers. The only way to handle these conveniently is to use powers of 10. A few examples are given below:

2×10^{10} = 20 000 000 000, i.e. twenty thousand million
3.54×10^{12} = 3 540 000 000 000
1.6×10^{-19} = 0.000 000 000 000 000 000 16

Very often one is only interested in approximate values, so that only one significant figure is quoted. For example, if I say that the number of electrons in a particular conductor at room temperature is 2×10^{24} m^{-3} you will know that it is about that value – it could be anywhere between 1.5×10^{24} and 2.5×10^{24}, i.e. two million million million million plus or minus 25% in this case.

If I want to write ten thousand million I should strictly write 1×10^{10}, but nobody ever does; they write simply 10^{10}, so if you see a power of 10 without a number in front of it you take the multiplier as 1.

A power of 10 without a multiplier is also used to indicate 'order of magnitude' which is even less accurate than one significant figure. Thus, the packing density of copper atoms in pure solid copper is often quoted as '10^{29} atoms per cubic metre' whereas a more accurate figure is 0.9×10^{29} or 9×10^{28} (the same thing).

1.2 MATHEMATICAL FUNCTIONS

I have tried to keep the maths to the minimum, but you will need to understand the ideas of differentiation and integration, and the subject involves a lot of exponential functions, which I will discuss briefly now.

Exponentials

There is a number which mathematicians call 'e' and which, like π, cannot be expressed exactly as a decimal (its approximate value is 2.718 28). This number appears in a lot of formulae, raised to various powers, and you will probably find on your calculator a key marked 'e^x'. An example of the use that I shall have to make of it is in the formula for the current in a semiconductor diode:

$$I = I_\mathrm{S}(\mathrm{e}^{\frac{eV}{kT}} - 1)$$

Unfortunately, the two 'e's used when the formula is written as above mean two different things: the e in the expression 'eV/kT' means the electronic charge. I could use a different letter for electronic charge, but that would be against all normal practice, so instead I shall do the same as other writers do when faced with this predicament and write 'exp(eV/kT)' to represent the exponential. The formula thus appears as

$$I = I_S[\exp(eV/kT)-1]$$

Note, in particular the following properties of exponentials:

$$\exp(x) \times \exp(y) = \exp(x + y)$$

$$\frac{\exp(x)}{\exp(y)} = \exp(x - y)$$

$$\frac{1}{\exp(x)} = \exp(-x)$$

and the inverse function:

if $y = \exp(x)$ then $x = \ln(y)$

where 'ln' is the 'natural' logarithm (you will also find this function on a scientific calculator).

1.3 UNITS

If, in a calculation, all the quantities are in SI units, then I shall not put the units into the calculation: for example, if a resistor of 100 Ω has a voltage of 5 V across it, I write the calculation of the current which flows as

$$I = V/R = 5/100 = 0.05 \text{ A}$$

On the other hand, I might write the calculation of the current produced by a voltage of 10 mV across 2 kΩ by

$$I = V/R = 10 \text{ mV}/2 \text{ k}\Omega = 5 \text{ } \mu\text{A}$$

or as

$$I = V/R = 10 \times 10^{-3}/2000 = 5 \times 10^{-6} \text{ A}$$

depending on the circumstance.

Where a unit is first introduced, or where it may not be clear what the unit is, the unit name is written out in full.

1.4 ROOM TEMPERATURE

Some of the things I shall calculate are very temperature-sensitive, so I must decide what I am going to take as 'room temperature'. The outside ambient temperature can easily vary between 270 and 300 K over a year. On the other

hand, most living and working rooms nowadays are maintained at about 294 K
(which is about 21 °C). For certain definitions in electronics, particularly to do
with noise, a standard value of 290 K is adopted and I shall use this value even
though it is rather lower than the temperature of some environments in which
the material may be used. If in a calculation I do not quote a temperature you
are to assume room temperature.

1.5 VALUES OF PHYSICAL CONSTANTS AND OF CERTAIN SEMICONDUCTOR PARAMETERS

These are listed in Tables 1.1 and 1.2. I have quoted the physical constants to
three significant figure accuracy. The semiconductor parameters are quoted to
an accuracy which is appropriate, bearing in mind that some of them are sensitive to temperature and to the chemical and structural purity of the material.

Table 1.1

Physical constants

Speed of light in a vacuum	c	3.00×10^8 m s^{-1}
Permittivity of a vacuum	ϵ_0	8.85×10^{-12} F m^{-1}
Electronic charge magnitude	e	1.60×10^{-19} C
Rest mass of the electron	m_e	9.11×10^{-31} kg
Planck's constant	h	6.63×10^{-34} J s
		$(4.14 \times 10^{-15}$ eV s$)$
Boltzmann's constant	k	1.38×10^{-23} J K^{-1}
		$(8.62 \times 10^{-5}$ eV K$^{-1})$

Note: at room temperature $kT/e = 0.025$ V, $kT = 0.025$ eV.

Table 1.2   Semiconductor parameters at around room temperature

	E_g^* (eV)	n_i (m^{-3})	μ_n (m^{2}V^{-1}s^{-1})	μ_p	D_n (m^{2}s^{-1})	D_p
Silicon	1.12	10^{16}	0.15	0.045	0.0039	0.0012
Germanium	0.67	10^{19}	0.39	0.19	0.010	0.0049
Gallium arsenide	1.42	10^{12}	0.85	0.04	0.022	0.0010
Indium phosphide	1.35	10^{14}	0.46	0.015	0.012	0.000 39

* E_g is given in electron volts (eV), the rest are in SI units.

2 Conduction in metals

In your studies up to now you have probably taken conductors and insulators for granted: electrons can flow in some materials when a voltage is applied, but not in others. In order to understand how semiconductor devices work it will be necessary to look a little more carefully into what is involved in the flow of an electric current. To do so we shall need to use the concepts of electric flux density and electric field, as well as voltage, and to apply some of the principles of electrostatics.

2.1 INTERACTION OF ELECTRIC CHARGES

Fundamentally, electric charge is a property of certain sub-atomic particles. The ones that will interest us are protons, each with one unit of positive charge, and electrons with one unit of negative charge. The magnitude of this unit of charge is approximately 1.6×10^{-19} coulombs, and it is often represented by the symbol e.

The effect of a charge in the space surrounding it can conveniently be represented as a flux, outwards from a positive charge and inwards to a negative charge, and the density of this flux, that is the flux per unit area perpendicular to its direction, is represented by a vector, D.

Charged particles are affected by the flux of other charges: positive charges experience a force in the direction of the flux, and negative charges in the direction opposite to the flux. The magnitude of the force per unit charge depends on the flux density and is represented by a vector E, in the direction of the flux, known as the electric field strength.

E and D are related by permittivity, which in free space is given the symbol ϵ_0, and which has a value 8.85×10^{-12} farads per metre. The relationship is

$$D = \epsilon_0 E$$

I can best illustrate the use of these electrical quantities by working out the force between a proton and an electron which are 10^{-10} m apart (about the radius of an atom; see Fig. 2.1). Notice that I have drawn the proton and electron as little spheres, but nobody knows how big they are, or even if they have a diameter in the normal sense. It is best to think of them as points. The arrows

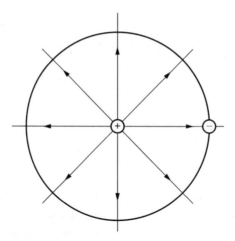

Figure 2.1   Electron in the field of a proton.

in Fig. 2.1 represent the flux of charge of the proton; they also indicate the direction of D and E.

The surface area of a sphere centred on the proton with the electron at its surface is $4\pi r^2$, i.e. in this case $4\pi \times 10^{-20}$ m².

The flux which passes uniformly through this area from the proton is 1.6×10^{-19} coulombs (the same value as the charge), so the value of D is:

$1.6 \times 10^{-19}/4\pi \times 10^{-20} = 1.27$ coulombs per square metre

The electric field produced by this flux is D/ϵ_0:

$E = 1.27/8.85 \times 10^{-12} = 1.44 \times 10^{11}$ volts per metre

And so the force on the electron is $E \times e$, and is towards the proton, since the electron is negative and experiences a force in the opposite direction to E:

$F = 1.44 \times 10^{11} \times 1.6 \times 10^{-19} = 2.3 \times 10^{-8}$ newtons

Self assessment test 2.1

Work through the above procedure using symbols instead of numbers.

Solution

You should arrive at the well known formula:

$$F = \frac{e^2}{4\pi\epsilon_0 r^2}$$

If I want to calculate the force on the proton due to the field of the electron I must start with Fig. 2.2. A similar calculation yields the same result, the force

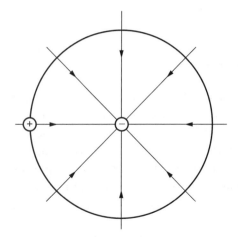

Figure 2.2   Proton in a field of an electron.

on the proton being towards the electron, as one would expect – the particles attract each other.

In general, if there are many charged particles, you can find the force on any one by working out the resultant flux due to all the others, adding the fluxes vectorially and deducing from the resultant the value and direction of E.

The fact that fluxes add vectorially relates to a very important principle known as Gauss's theorem. This states that for any volume, the total electric flux emerging from that volume is equal to the net electric charge inside the volume. Flux may be directed in and flux may be directed out, and if the total outward flux is equal to the total inward flux then there is no net charge inside. There may be positive charges and negative charges, but there must be equal amounts of both. If more flux is directed out than is directed in, the net charge inside is positive; if more flux is directed in than out the net charge inside is negative.

Since the proton and the electron that we have been discussing each experiences a force towards the other, work would have to be done, and hence energy put in, to move them apart. I can calculate the work required to move the electron to infinity while holding the proton fixed. Since the force will not be constant but will decrease as the electron is moved, an integration is required: $\int_r^\infty F \, ds$, where s is distance.

Self assessment test 2.2

The integral can be written $\dfrac{e^2}{4\pi\epsilon_0} \displaystyle\int_r^\infty \dfrac{1}{s^2} ds$. Carry out the integration.

Solution

You should get $\dfrac{e^2}{4\pi\epsilon_0 r}$.

This works out, for the initial separation of 10^{-10} m, as 2.3×10^{-18} joules.

You will have noticed, if you did not already know, that the units of E are volts per metre, and this indicates that voltage, which is force per unit charge times distance, measures work per unit charge, and thus energy change per unit charge (a fact which is not always obvious when one is using voltage in a circuit context). Quantities of energy involved in semiconductor theory are usually very small values in joules, so it is convenient to use a smaller unit, the electron volt (eV).

One electron volt is the energy change which occurs when an electron moves through a potential difference of one volt, and so its value in joules is $e \times 1$, i.e. 1.6×10^{-19} J.

Measured in electron volts, the work done in moving the electron from 10^{-10} m distance from the proton to infinity is

$2.3 \times 10^{-18}/1.6 \times 10^{-19} = 14.4$ eV

If I started with the electron 2×10^{-10} m from the proton, the energy required to remove it would be 7.2 eV; the energy to remove the electron is inversely proportional to initial separation, as can be seen from the formula above.

2.2   PROPERTIES AND ELECTRICAL BEHAVIOUR OF COPPER

The nucleus of a copper atom contains 29 protons. Viewed from a distance the copper atom has no resultant electric flux, and so appears electrically neutral, because the nucleus is surrounded by 29 electrons. The separation of these electrons from the nucleus is very large compared to the nuclear diameter, and so it is natural to wonder why the electrons do not all crowd in as close as possible to the nucleus. The electrons repel each other, but this does not explain their arrangement. At one time it was thought that the electrons must orbit the nucleus, like planets round the sun, so that the electrical attraction to the nucleus would be counterbalanced by the centrifugal effect, but this is not an adequate explanation either. To gain a better understanding of the behaviour of electrons in atoms you would need to study a theory known as quantum mechanics which is outside the scope of this book, although I shall quote some of its results from time to time.

In fact we have no way of knowing exactly what the electrons in the atom are doing at any one time, but we can measure the energy required to remove an electron, which I shall now refer to as the binding energy, and from such measurements it emerges that the electrons all have specific binding energies of which no more than two can be the same. It is conventional to take the energy of an isolated electron, at rest, as zero, and with this reference the electric field energies of the electrons in the atom are all negative, since work must be done to remove them. Some of an electron's energy may be positive kinetic energy, since it seems likely that the electrons move about in the atom, but the net energies are fixed and negative. Those electrons with the largest binding (negative) energies are nearest the nucleus.

Of the copper atom's electrons, one is much more loosely bound than the rest (has a much smaller binding energy) and is much further from the nucleus. It is thus convenient for many purposes to consider the nucleus and its 28 inner electrons as one particle, which we can describe as a copper ion with a charge $+e$, accompanied by this single loosely bound electron, which is often termed a 'valence' electron because it is responsible for the chemical behaviour of copper.

I want now to discuss the structure of solid copper, but in order to do so I shall imagine that the solid is formed by cooling a copper gas first to form a liquid, and then as it is cooled further to form the solid. This will allow me to emphasize the important role of heat in the properties, including the electrical behaviour, of solids at room temperature.

Heat energy in a substance is random kinetic energy of its constituents: atoms, molecules or, as we shall see, 'free' electrons. The temperature is a measure of the average energy per particle. The constituent particles of a substance are continuously exchanging heat energy by collisions and other interactions, so that at any time there will be a spread of heat energy among the particles, but with an average value per particle related to the temperature.

The atoms of copper gas at a high temperature will be moving at high speeds and will collide with such violence that they will fly apart again. As heat is lost to the surroundings and the gas cools, atoms which come into contact tend to adhere because the valence electrons have time to arrange themselves so as to form an electrical bridge holding together the positive ions – this is stable because it represents a lower energy condition than the isolated atoms. The atoms have too much energy to stay in fixed positions and still move relative to each other in this liquid form. As further heat is lost the ions settle into a regular matrix (a 'crystal structure' in fact) and the heat energy of the ions now becomes oscillations about fixed positions. In this state the ions are so close together that a valence electron is as near to the neighbouring ions as to its original parent ion and is no longer attached to one ion. These electrons are effectively free particles within the solid and, taking a share of the thermal energy, move around rapidly within the solid.

It is interesting to try to estimate the average speed of 'free' electrons in copper at 290 K (room temperature). If the electrons were truly free particles in a vacuum their average kinetic energy would be $3kT/2$, where T is the absolute temperature and k is Boltzmann's constant, of value 1.38×10^{-23} joules per Kelvin (JK^{-1}).

The mass of an electron is 9.11×10^{-31} kg, so if I take

$$\frac{1}{2}mv^2 = \frac{3}{2}kT$$

$$v = \sqrt{(3kT/m)} = \sqrt{(3 \times 1.38 \times 10^{-23} \times 290/9.11 \times 10^{-31})}$$

$$\approx 10^5 \text{ m s}^{-1}$$

i.e. about one hundred thousand metres per second.

In fact, much of the valence electrons' motion is not due to heat. If the copper could be cooled to absolute zero temperature, so that the ions had no vibrations, the valence electrons would still have a range of energies, and thus motions, associated with their function of moving between the ions to hold them together.

The picture that emerges of a piece of copper at absolute zero temperature is of a regularly spaced matrix of copper ions – nuclei with tightly bound electrons – with valence electrons, one per ion, moving in individual ways throughout the structure and performing a complex 'dance' such that there is always an average of negative charge between the positive ions, holding them together, and the instantaneous distribution of the electrons is always uniform so that any significant volume of the copper is at all times electrically neutral. At room temperature, the ions vibrate and the motions of the valence electrons are perturbed by heat. These electrons still hold the structure together effectively, but, as a result of the perturbation, small transient departures from electrical neutrality occur, resulting in the phenomenon of electrical noise.

From here on when I want to refer to valence electrons in copper I shall simply refer to them as 'electrons' if there can be no confusion with the electrons bound in the ions.

Because the electrons are effectively 'free' in a piece of copper it is possible to add a few more so that the piece is negatively charged. The rest of the electrons repel the extra ones so that they form a layer of negative charge on the surface. Similarly a few electrons can be removed leaving the piece charged positive, and again the charge resides on the surface by virtue of the fact that electrons move back to expose the charges of the ions at the surface.

A capacitor has two plates which can be charged in this way. It will be very useful to me later if I discuss now the distribution of flux, field and voltage in a parallel plate capacitor.

Figure 2.3 shows a parallel plate capacitor made of two copper sheets: one has been charged positive and the other one equally negative, perhaps by connecting the plates to the terminals of a battery (the plates will always be equally charged). All the charge resides on the faces of the plates closest to each other, because charges on the two plates attract each other. Each

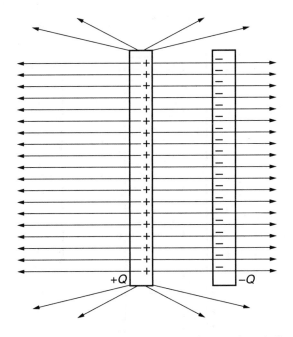

Figure 2.3   A parallel plate capacitor, with plates of area *A* and distance apart *d*. The electric flux shown is that due to the positive charges only.

individual electron or exposed ion will give rise to a sphere of flux, but the resultant of these overlapping spheres of flux will be perpendicular to the plates, except at the ends. I have just shown the flux due to the positive charges in the diagram. If I ignore the ends, which I can do if the plates are large and close together, then the flux passing from the positive plate through the negative plate is, in total, $Q/2$. Putting the flux of the negative charges on the diagram would overcomplicate it, but you should be able to see that the fluxes would overlap, and since the flux of the negative charges will be in towards the negative plate, the two sets of flux will reinforce between the plates and cancel outside them. The total flux between the plates is $Q/2 + Q/2$, i.e. Q, and the resultant flux appears to terminate on the charges at both ends, as shown in Fig. 2.4. The flux density between the plates is Q/A.

   If the space between the plates contains a material other than a vacuum, then to find the value of E, D must be divided by a permittivity value different from ϵ_0. The charges in the atoms of the material are displaced by the flux from the plates and alter the net flux value. Fortunately, however, we can avoid this complication by ascribing to the material a 'relative permittivity', a number which amends the relationship between the flux from the plates and the field to

$E = D/\epsilon,$

where $\epsilon = \epsilon_0\epsilon_r.$

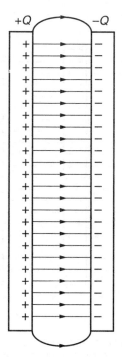

Figure 2.4   A parallel plate capacitor, with plates of area *A* and distance apart *d*. The resultant electric flux due to both sets of charges is shown.

Self assessment test 2.3

Deduce the electric field, and from this the voltage between the plates, and hence derive the formula for the capacitance of the two plates, $C = \epsilon A/d$

Solution

Since, ignoring edge effects, the flux density between the plates has a uniform value of Q/A, $E = D/\epsilon = Q/\epsilon A$.

As D does not vary with distance between the plates, E is constant, so the voltage between the plates must be $E \times d$, $= Qd/\epsilon A$.

$C = Q/V$ so, in this case, $Q/V = Q/(Qd/\epsilon A) = \epsilon A/d$.

Suppose that one tries to pass an electric field through a piece of copper. An obvious way to attempt this is to put the piece of copper between the plates of a charged capacitor as shown in Fig. 2.5. The electric flux between the plates of the capacitor will exert a force on the electrons in the copper and, quickly, layers of charge will establish themselves on the surface of the copper. The flux due to these layers of charge on the copper surface exactly cancels the applied flux within the copper; if it did not do so then the remaining field would continue to move electrons until it did. It is impossible to pass an electric field

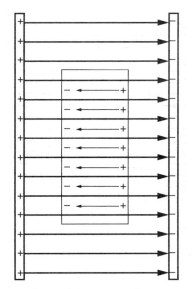

Figure 2.5   Induced charge and flux inside a piece of copper in an electric field.

through a conductor in this way, although, as we shall see, a long-range electric field (as distinct from the short-range fields within the ions) exists in a conductor which is carrying an electric current. Applying the notion of relative permittivity here would suggest that one must ascribe to the copper an infinite value of ϵ_r. However, when a current flows in copper it is associated with an electric field and a flux density, and the relationship between the two is influenced by the displacement of bound charges in the ions. Thus, by convention, a value of relative permittivity ascribed to a conducting material excludes the effect of free charges.

Practical experience shows that when the free electrons rearrange themselves to cancel an external flux the resulting perturbation is not sufficient to influence significantly their function of binding the ions. I can demonstrate this by putting in some numbers. The density of copper atoms in solid copper is about 10^{29} per cubic metre and there are the same number of free electrons. A 1 cm cube of copper will thus contain 10^{23} atoms. A monatomic layer on one face of a 1 cm cube will contain

$$(10^{23})^{2/3} = 2.15 \times 10^{15} \text{ copper ions}$$

and this will also be the number of electrons in the undisturbed layer. Suppose an external flux that would produce a field of $10^4 \, \mathrm{V\,m^{-1}}$ in free space (equivalent to 100 V across the cube) is applied perpendicular to a pair of faces of such a cube – how many electrons are transferred from one face to the other and how large a disturbance of the surface density of electrons does this represent?

This problem will allow me to introduce some ideas that will be very useful later (see Fig. 2.6). I shall neglect the small amount of flux terminating on the side faces of the cube. The surface layers of charge on the perpendicular faces

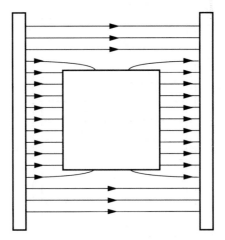

Figure 2.6   A cube of copper in an electric field.

must equal the flux arriving at those faces; we can conveniently say that the flux terminates on the surface. I can calculate the incident flux density:

$$D = \epsilon_0 E = 8.85 \times 10^{-12} \times 10^4 = 8.85 \times 10^{-8} \, \text{C m}^{-2}$$

The total flux over an area of 1 cm², i.e. 10^{-4} m², is

$$8.85 \times 10^{-8} \times 10^{-4} = 8.85 \times 10^{-12} \, \text{C}$$

This equals the required surface charge, so the number of electrons added or removed is

$$8.85 \times 10^{-12}/1.6 \times 10^{-19} = 5.5 \times 10^7 \, \text{electrons}$$

This is a very tiny fraction of the total number of electrons in the surface layer.

Self assessment test 2.4

Since the free electrons in copper are moving around, some at high speed, why do they not escape from the solid altogether?

Explanation

Inside the copper the electrons move freely (subject to the proviso that the overall charge equilibrium is not disturbed), but if an electron leaves the surface then the material in the vicinity is minus an electron and thus positively charged, so electric forces pull the electron back.

In fact, some of the more energetic electrons constantly emerge from the surface and then return, so that there is a cloud of electrons just outside the surface. This phenomenon is exploited to obtain thermionic emission in devices such as the cathode of a television tube (using a more appropriate material than copper).

The energy needed for an electron to escape altogether is known as the work function of the surface. Giving the material a negative charge makes the work function smaller and a positive charge makes it larger, as you might expect.

2.3   METALLIC CONDUCTION

An electric current in a metal wire consists of a flow of 'carrier' electrons through the wire; these carrier electrons are the same 'free' electrons that we have been discussing. Suppose a wire, carrying a current I amperes (A), has a cross-sectional area A m². Assume that the average velocity of the electrons along the wire is v m s⁻¹. This velocity is superimposed on the random motions of the electrons and differs from them in that it is organized in one direction. Now consider a length of the wire that is the same length in metres as the average electron velocity in metres per second – see Fig. 2.7. If the carrier density, that is the number of electrons per cubic metre, is n, then the number of carriers in this length of wire is nAv. The total charge carried by all these electrons is $nAve$ coulombs. In one second, the amount of charge which passes the plane marked P in the diagram (or any other plant perpendicular to the wire) is $nAve$ coulombs; but this is the definition of the current in amperes in the wire, so

$I = nAve$ A

The result can be made independent of the dimensions of the wire by dividing by A. I divided by A is called the current density, J, and so

$J = nev$ A m⁻²

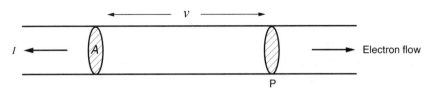

Figure 2.7

Self assessment test 2.5 _____

Calculate the velocity v (as above) for electrons in a copper wire of cross-sectional area 1 mm² carrying a current of 1 A.

Solution

$1 \text{ mm}^2 = 10^{-6} \text{ m}^2$, therefore

$$v = I/(nAe) = 1/(10^{29} \times 10^{-6} \times 1.6 \times 10^{-19})$$

$$= 6.25 \times 10^{-5} \text{ m s}^{-1} = 22.5 \text{ cm per hour}$$

which is a very small velocity compared to some of the random electron velocities!

Now we know from circuit theory that a voltage applied between the ends of a conductor produces a current and that the current is proportional to the voltage, the two being related by resistance or conductance. For convenience, here I am going to work with conductance. Ohm's law, then, can be stated as

$I = GV$, so $G = I/V$

The conductance is the property of a specific component. To describe the property of a material I need conductivity, σ.

Self assessment test 2.6

You are probably familiar with the relationship between resistance and resistivity, $R = \rho L/A$. Deduce from this the relationship between conductivity and conductance.

Solution

$$\frac{1}{G} = \frac{1}{\sigma} \times \frac{L}{A} \qquad \text{so } \sigma = G\frac{L}{A}$$

This means that

$$\sigma = \frac{I}{V} \times \frac{L}{A}$$

But I/A is J, and V/L is E, so $\sigma = J/E$ or $J = \sigma E$.

We have already seen that for any current, the current density $J = nev$; in this case I shall write the relationship as

$$J = nev_d$$

v_d is called 'drift velocity' and is the (rather inappropriately named) velocity of current carriers which can be attributed to a voltage, and thus to the presence of an electric field (there is another mechanism, diffusion, which can also cause current carriers to move – this is described later in the book). So

$$nev_d = \sigma E \qquad \text{or } \sigma = ne(v_d/E)$$

The quantity v_d/E, that is the drift velocity per unit electric field, is considered to be a property of the electrons in the particular material and is known as the electron mobility, μ_n (units: metres squared per volt second, $m^2\,V^{-1}\,s^{-1}$).

Finally, then, I can write

$\sigma = ne\mu_n$ siemens per metre

From the above it should be clear that Ohm's law, which may be regarded as an experimental observation, indicates that an electric field in a metallic conductor causes the carrier electrons to move with a steady drift velocity proportional to the magnitude of the field. Two questions arise: what is implied by the fact that the electrons move with a steady drift velocity, and how does the electric field get into the conductor?

Steady carrier drift velocity

Electrons in free space, when subjected to an electric field, accelerate. The most familiar example of this is in a television tube: the electric field in the electron gun continuously increases the kinetic energy of the electrons and this energy is given up suddenly when they hit the screen. In a conductor the fact that the electrons do not accelerate indicates that there is a continuous mechanism which extracts energy from the electrons as they move down the potential gradient: this energy generally emerges as heat. The mechanism is some sort of interaction between the electrons and the array of ions and we know that this is temperature-dependent, since the resistance of a metallic conductor increases with increasing temperature. One can imagine that the electrons bump into the ions continuously as they move, and that such collisions are more likely as the temperature rises because the increasing vibration of an ion will give it a larger collision cross-section. Of course the electrons frequently bump into ions on account of their thermal motion as well, but that is all part of the continuous exchange of thermal energy between particles and does not result in an increase in the total heat content. The notion of electrons 'bumping' into ions is naive, but clearly there is some sort of interaction, and this gives a simple mental picture.

The electric field in a conductor due to an applied voltage

If I take a battery, or a d.c. generator, I know that it will give me a certain terminal voltage, but I also know that a voltage difference implies an electric field – so how does this work in a circuit?

Generators are sources of emf (electromotive force), and it is worth pausing for a moment and asking what they actually do. Take a battery for instance. A chemical process occurs inside the battery which causes one plate to have an excess of electrons and the other to have too few. Irrespective of the size of the battery, for a given type, the voltage, that is the available energy per unit charge transferred from one terminal to the other, is fixed. If the terminals of the battery are connected together externally (ignoring the fact that in practice there is a limit to the size of the current you can draw) the electrons flow out of the negative plate and into the positive one and the chemistry in the cell works continually to restore and maintain the difference, until it is exhausted. In a generator, a conducting coil moves in a magnetic field, pushing electrons in the coil so that the density in one terminal is enhanced and in the other depleted, resulting in terminal characteristics similar to those of the battery. A source of emf, then, is an electron pump.

What happens in a wire connecting the terminals of a source of emf? The voltage drops uniformly along the wire irrespective of its shape; it does not have to be stretched out in a straight line. This suggests that the charge distribution which must be associated with the electric flux producing the field must be all along the wire. The flux which causes the field which drives the electrons through the wire is produced by the distribution of the electrons themselves.

It is often asserted that in a conductor carrying a current there is charge neutrality – in other words, the electron density everywhere cancels out the ion density just as when no current flows – but this cannot be precisely true. Let me first show where this idea comes from. The flow of current in a circuit is the same at all points: this is Kirchhoff's first law. In a uniform conductor the flowing electrons will be travelling at the same speed at all points and so giving up energy at the same rate at all points. Thus the loss of energy per unit length, and hence the drop of potential per unit length, will be constant. A uniform potential drop implies a uniform electric field along the direction of current flow. In turn this implies a constant value of D along the direction of current flow.

Look at Fig. 2.8. A constant value of D implies the same flux into the volume V as out of it, and so, from Gauss's theorem, no net charge in the volume V *provided no flux leaves the side of the wire*. But in practice some flux will leave the wire, particularly if the wire is thin and if it twists and turns. The charge distribution will adjust itself so as to maintain at all points in a uniform wire a constant value of the electric field strength along the direction of the current. It is commonly the case in electric circuits that energy distribution dictates voltage distribution, and charges arrange themselves accordingly.

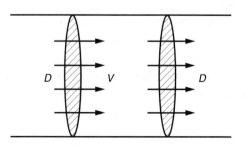

Figure 2.8

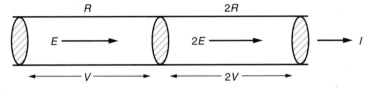

Figure 2.9

Consider now two equal lengths of wire of the same diameter connected in series, the materials of the two having different resistivities so that the second has twice the resistance of the first, as shown in Fig. 2.9. With a current flowing, the voltage drop down the second piece of wire will be twice that down the first, so the value of longitudinal E in the second wire will be twice that in the first. Application of Gauss's theorem at the junction indicates that there must be a net charge at that point.

[*Note*   If the lengths of wire are made of different metals there may be a double layer of charges of opposite signs at the junction, irrespective of any current, due to a difference in the average electron energies in the two metals. This leads to a 'contact potential' which I shall discuss later in the context of semiconductors. I am not concerned with this effect here.]

2.4 SUMMARY

In a simple solid metallic conductor, of which copper is the most appropriate example, the crystal structure is a matrix of positive ions held together by an assembly of electrons, equal in number to the ions, which move between the ions in a way which tends to maintain charge neutrality at all points. With this last proviso the electrons are effectively free within the solid and not linked to individual ions. The electrons are in thermodynamic equilibrium with the lattice and so their motions are perturbed by heat, however, the only immediate effects of this are second order – electrical noise and thermionic emission.

Any assembly of charges near a piece of metallic conductor causes the electrons to organize themselves on the surface of the conductor so as to terminate the incident flux and maintain charge neutrality within.

In an electric circuit a voltage is normally applied by a source of emf which injects electrons into one end of the circuit and removes them from the other. Energy considerations within the components of the circuit determine the voltage distribution in the circuit: the charges in the conductors arrange themselves in such a way as to set up the fields dictated by the voltage distribution. Under these circumstances there may be a departure from charge neutrality in parts of the conductors.

For any current of electrons in a solid one can write for the current density

$J = nev$

If the current is caused by an electric field, then this equation can be rewritten

$J = nev_d$

v_d being drift velocity. In a metal the drift velocity has a very small value, for normal current magnitudes, and is an organized movement of charge superimposed on very much larger random movements.

The conductivity of a metal can be expressed as

$\sigma = ne\mu_n$

The mobility of the electrons, μ_n, is defined as the drift velocity per unit electric field. It is a property of the material rather than of the electrons, and represents the retarding effect of the ion matrix on the electrons' drift component of motion. There will be a quoted value for μ_n in copper and another for μ_n in aluminium, say, but even for a given material, the value varies with material purity, mechanical defects in the crystal and temperature.

2.5 PROBLEMS

1 Two parallel sheets of metal have a block of metal between them, with dimensions shown in Fig. 2.10. Calculate the approximate capacitance between the two sheets.

2 Two wires each of the same thickness (1 mm^2 cross-sectional area) are joined together. The material of one wire has a conductivity of 2×10^4 S m^{-1}

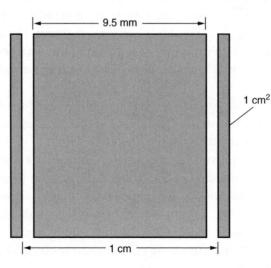

9.5 mm

1 cm^2

1 cm

Figure 2.10

and the other twice this value. The relative permittivities of the materials of the wires each have a value of 5.5. There is no contact potential at the junction. If the wires carry a current of 1 A, calculate the magnitude of the charge at the junction. How many electronic charges does this represent? What determines the sign of the charge?

3 The resistivity of copper is quoted as 10^{-8} Ω m. From this figure, what is the value of the electron mobility in copper?

4 A particular material (not a metal) contains carrier electrons with a density of 10^{23} m^{-3} which allows it to conduct electricity. The electron mobility in the material is 0.15 m^2 V^{-1} s^{-1}. What is the conductivity of the material, and what is the drift velocity of the electrons when a current of 100 mA flows in a cross-sectional area of 1 mm^2?

3 Conduction in intrinsic semiconductors

The term 'intrinsic', as used here, means that the material is as pure as it is practically possible to make it. In the next chapter I shall introduce 'extrinsic' semiconductor materials to which, after purification, measured amounts of certain other elements have been added.

By definition, a semiconductor is a material with a conductivity intermediate between that of a good conductor, like copper, and an insulator, like polystyrene. From the equation

$\sigma = ne\mu$

You can see that the lower conductivity may be a result of a smaller carrier density (n) or a lower mobility (μ), or both. In fact, in the materials that we shall be interested in, the mobility is higher than in copper; the lower conductivity is a consequence of there being far fewer carriers per unit volume.

Apart from having a carrier density of appropriate value for certain purposes, the other important property of the semiconductor materials used in electronics is that, in them, another sort of current carrier besides electrons exists. These second types of carrier have a positive charge and are known as positive holes.

3.1 THE PROPERTIES AND ELECTRICAL BEHAVIOUR OF PURE SINGLE-CRYSTAL SILICON

Silicon is by far the most commonly used semiconductor material in electronics, so I shall develop the theory taking it as my example.

Silicon forms crystals in a way which is different and more complicated than is the case with copper. In copper the ions behave like spheres and form regular layers like a pile of cannon balls – that is, by the spheres packing together as closely as possible in a layer and the next layer sitting in the indentations formed by sets of three in the layer below. This is the closest way of packing spheres together: the free electrons hold the ions together but do not determine the form of the crystal.

The silicon atoms in a silicon crystal are held together by what are called covalent bonds. In a covalent bond a pair of valence electrons is shared by two atoms, one being provided by each atom. Again it is the negative charge of the valence electrons which holds the positive 'cores' (nucleus plus inner electrons) together, but these valence electrons are confined to a particular small region with respect to the atoms and cannot move – in the way that the free electrons in copper do – under the influence of an electric field. A silicon atom has four valence electrons, and the crystal forms with each atom attached symmetrically to four neighbours. The arrangement cannot be drawn on a flat diagram. I've tried to draw it in perspective in Fig. 3.1, but the atom cores are shown relatively much further apart than they actually are. It is difficult to imagine the way that this structure fits together in three dimensions without constructing a model. Although the silicon atoms are close together, their configuration does not give such close packing as is the case for copper, and so there are 5×10^{28} silicon atoms per m^3 of silicon crystal as opposed to 9×10^{28} copper atoms per m^3 in a copper crystal (i.e. little more than half as many).

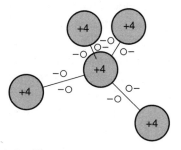

Figure 3.1   Valence bonds in silicon.

There is no way of knowing exactly how the valence electrons move within the regions that they occupy; it is generally assumed that they continuously exchange places throughout the crystal, but certainly they do not take part in electrical conduction – if the temperature is near absolute zero. The reason for this last proviso is that at room temperature (and to a diminishing extent at temperatures below this) the vibration of the atoms due to heat causes some of the valence electrons to be shaken free and to become conduction electrons, free to move around inside the crystal in the same way as do the free electrons in copper.

The gaps in the valence structure, left by the electrons which are shaken free, are positive, in the sense that each represents a missing negative charge from a system which overall is electrically neutral. Valence electrons seem to be able to move freely from one valence position to another (hence the assumption that they exchange places), so that the gaps, or 'positive holes', propagate throughout the structure by the combined movement of the valence electrons, as if they also were free (positive) particles. The existence of the free electrons and positive holes explains why silicon is a semiconductor at room temperature.

The nature of conduction electrons and positive holes in silicon

Where there is no confusion I shall refer to positive holes simply as 'holes'.

Electrons are strange things which are not 'like' anything that we can see or touch, and some of their properties seem very odd if you try to think of them as specks of dust or something of that sort. For most purposes I can model them as conventional particles, with a negative charge and a tiny mass; but always remember that this representation is incomplete. In particular, to analyse how a conduction electron moves in a crystal of silicon (or any other conducting solid) under the influence of an applied electric field, I can consider just the force on the electron due to the applied field together with some sort of 'frictional' force resisting its motion. Such an analysis, however, ignores the detailed structure of the solid with its internal electric fields in and around the core atoms, and it also ignores certain 'wave-like' properties of electrons. These complications which have been ignored can be compensated for by taking a figure for the mass of the electron which is different from the mass as measured in free space: the result is called the effective mass of the electron, symbol m^*. Unfortunately, there is not one value for the effective mass of electrons: it depends on the nature and condition of the conducting material and on the temperature (it may even depend on the direction in which the electron is moving in the crystal). The effective mass of electrons in a solid under given conditions can be measured experimentally.

A positive hole is a place in the valence structure of a solid where an electron is missing. Such holes are found to move under the influence of electric fields and to have their motions perturbed by heat in the same way as conduction electrons. The motion of a positive hole is a consequence of the incredibly complex movement of all the valence electrons in the solid within the valence regions, and yet it turns out that positive holes have an effective mass which can be measured and is found to be positive (not negative, as you might expect for a place where something is missing) – in other words they can be modelled as conventional positively charged particles. Again the value of the effective mass of holes depends on the material and its condition and the temperature. For a given material and conditions, the effective mass of holes is generally different from the effective mass of conduction electrons.

There is a link between effective mass and mobility; some semiconductor materials have a small electron effective mass and in these the electron mobility is high. The effective mass of holes does not vary so widely, and in general holes have lower mobility than electrons.

An energy band diagram for silicon

In order further to explain the electrical properties of pure silicon I am now going to introduce a sort of graph which is called an energy band diagram. This graph shows all the energies which are available to the valence electrons.

Initially I shall consider just one small region within a pure silicon crystal.

For this, the energy band diagram is a one-dimensional graph with just a vertical axis along which are marked all the possible energies which valence electrons can have in that region. In Fig. 3.2, the horizontal dimension has no significance other than to make the diagram easier to read. Remember that an electron's energy at any instant is part potential (depending on where it is relative to the charges around it) and part kinetic (depending on its motion) and that the total is negative relative to the 'vacuum level', which is the energy of an electron away from the crystal.

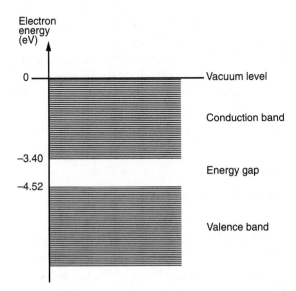

Figure 3.2   Energy band diagram for intrinsic silicon.

There is a band of energies called the valence band. This represents all the energies possessed by electrons which occupy the valence regions in the crystal and carry out the valence (crystal binding) function. From quantum mechanics we learn that within an interacting system only two electrons can have the same energy, and so the band in fact consists of a large number of closely spaced separate levels. The width of the band, between the highest and lowest levels, is determined by the dimensions of the crystal lattice (distance between atoms). The number of levels depends on how many atoms are involved; thus, if we consider one cubic millimetre of the crystal, that is 10^{-9} m^3, there will be $5 \times 10^{28}/10^9 = 5 \times 10^{19}$ atoms requiring four valence electrons each and one level per two electrons, giving 10^{20} separate valence levels. Because of the high density of levels I cannot draw them individually.

Above the valence band is a 'forbidden band' of energies which, in a pure crystal, no electron can possess, and above this is the conduction band, which extends up to the vacuum level. Qualitatively, I can explain this as follows. An electron can leave the valence band if it can move to regions of the crystal where it is further from the core atoms – in the 'spaces' in the crystal structure. Here it can move around with a range of velocities, exchanging thermal energy and moving under the influence of any external electric field. The move to these 'free electron' regions requires a step in potential energy which accounts for the energy gap. Again, the conduction band consists of densely packed individual levels.

The energies of the top of the valence band and the bottom of the conduction band are shown in Fig. 3.2: from these I can deduce two significant values. The first is the energy gap, E_g, which you will see is 1.12 eV. The second is the energy which an electron at the bottom of the conduction band needs to acquire to escape from the crystal; this is known as the electron affinity and is about 3.4 eV. Both these quantities vary slightly with temperature. The values given are for a silicon crystal at room temperature.

For a pure single crystal of silicon, the energy band diagram will be the same everywhere – indeed I could draw one diagram to represent the whole crystal, with an appropriate number of levels in the bands – except at the surface. The surface will generally have a layer of oxide or other impurities, but even if it is absolutely clean and in a vacuum, the discontinuity of bonding affects the energy levels.

Intrinsic carrier densities in silicon

To become a conduction electron, leaving behind a hole, a valence electron has to acquire at least 1.12 eV of energy and this it can only get from the heat in the crystal (except in those special cases where light energy is introduced). The heat energy in a material is constantly redistributed amongst the participating particles (those which are able to take up heat energy individually) by their interactions, so that, although at a given temperature the average energy per particle is constant (and at room temperature is much less than 1.12 eV), the individual energies vary with time and range widely about this average. When, by chance, a silicon atom acquires for a moment sufficient energy, one of its valence electrons may be shaken free and become an individual particle, now taking part in the distribution of heat energy. Strangely, the hole left behind also becomes a 'free' particle and also takes up heat energy. If this were the only mechanism, the number of carriers would continue to grow with time, but there is a counter effect whereby a hole and conduction electron meet and recombine, giving heat energy back to the lattice. The probability of recombination increases with increasing populations so that, at any given temperature, there is a dynamic equilibrium between creation and recombination, with a fixed *number* of conduction electrons and holes (although the individuals appear and disappear). It is actually possible to measure the 'carrier lifetime', that is, the average length of time that an individual conduction electron or hole exists. The equilibrium value of electron and hole populations (equal, naturally) increases rapidly with temperature. The number of electrons, and equally the number of holes, is called the intrinsic carrier density and given the symbol n_i, its value at room temperature is about 10^{16} m^{-3}.

Conductivity of pure silicon

Electrons and holes both contribute to the current when a voltage is applied. Electrons drift one way and holes the other, but the currents which they carry add to each other (the notional direction of the current being that in which the holes move). By an extension of the argument given in Chapter 2 the formula for conductivity is

$$\sigma = n_i e \mu_n + n_i e \mu_p = n_i e (\mu_n + \mu_p)$$

μ_n being electron mobility and μ_p hole mobility.

As temperature increases, the mobilities tend to fall, but this is completely swamped by a very rapid increase in intrinsic carrier density, so that conductivity rises very rapidly with temperature.

Self assessment test 3.1

Using values given in Tables 1.1 and 1.2 on page 4, calculate the conductivity of pure silicon at room temperature.

Answer

You should get the result $\sigma \approx 3 \times 10^{-4}\ \mathrm{S\ m^{-1}}$.

3.2   THE FERMI FUNCTION

In the case of dynamic equilibrium, at a fixed temperature, in which a proportion of electrons are constantly jumping to higher energy levels and others falling back to lower levels, it is possible to calculate for any given energy the probability that an electron will be at that energy – if a level exists there. The formula for this probability, called the 'Fermi function' after the man who worked it out, is

$$f(E) = \frac{1}{\exp[(E - E_F)/kT] + 1}$$

where k is Boltzmann's constant, T is the absolute temperature (Kelvin) and E is the energy of the level concerned. E_F is a special energy known as the 'Fermi level'. I shall now explain what this formula means and how it is applied.

Figure 3.3 shows a plot of the Fermi function for three different temperatures.

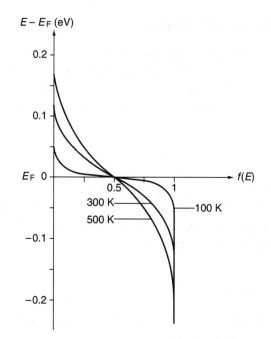

Figure 3.3   The Fermi function for different temperatures.

Self assessment test 3.2

Put some values into your calculator to check these curves. If you use the value of Boltzmann's constant, given in eV K^{-1} (see Table 1.1 page 4), you can use the values of $E - E_F$ in eV in the formula.

Result

When you have done this you should have a better feel for how the function works and you should realise that it is symmetrical about the Fermi level and approaches zero asymptotically at energies above E_F and unity asymptotically at energies below E_F.

Self assessment test 3.3

Think what the graph of the function would be like if the temperature were absolute zero.

Answer

You cannot use the formula directly because of the difficulty of dividing by zero, but it should be apparent that, for smaller and smaller values of T, the curve will get closer and closer to a condition where $f(E)$ is unity for energy values right up to E_F and zero for all values above it.

At the Fermi level, $f(E)$ is 0.5, so it is a sort of 'balance energy' about which the electron energies are perturbed. The Fermi level, however, is not fixed; its value changes with temperature and has to be determined by considering the distribution of the allowed energy levels in the material.

Perturbation of electron energies in copper

The energy band diagram for copper is very simple: valence electrons are conduction electrons, and there is just one band extending from the lowest energy which a conduction electron can possess up to the vacuum level. There are far more available levels in the band than there are electrons (simple theory suggests twice as many), and at absolute zero temperature the electrons will occupy the lowest available levels, so the probability of occupancy goes from zero to unity at the highest occupied level – this, then, is the Fermi level (see Fig. 3.4(a)). If you now imagine Fig. 3.4(a) modified by the probability of occupancy curve for 300 K shown in Fig. 3.3, the result will be as indicated in Fig. 3.4(b) – there will be some empty levels just below the Fermi level, and some occupied levels just above the Fermi level. The Fermi level will not change significantly.

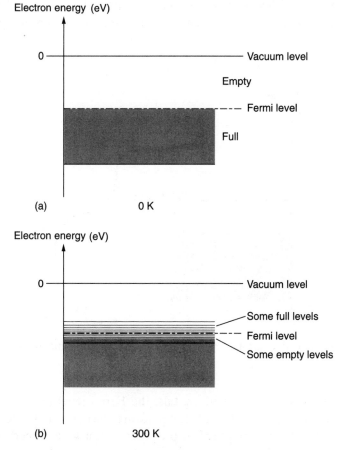

Figure 3.4 Perturbation with temperature of electron energies in copper.

[*Note*   At very high temperatures the Fermi level in a metal will be a little different from its low temperature value. This is caused by the density of energy levels not being uniform throughout the band in which conduction takes place.]

The properties of copper which I discussed in Chapter 2 are consistent with the energy band model given in Fig. 3.4. Since there are plenty of empty levels immediately adjacent to full levels in the band there is nothing to prevent the electrons taking on the small extra increments of kinetic energy associated with drift velocity in a current – as soon as one electron moves up another one can take its place, and effectively, all the free electrons can take part in conduction. Similarly the addition or subtraction of a few electrons, associated with charging a piece of copper, can readily be accommodated.

Perturbation of valence electron energies in pure silicon

At absolute zero temperature the valence band is full and the conduction band is empty: the Fermi level is at the top of the valence band. At any temperature above absolute zero there will be empty levels in the valence band and the same number of full levels in the conduction band; the balance between these indicates that the Fermi level must be in the middle of the energy gap (see Fig. 3.5).

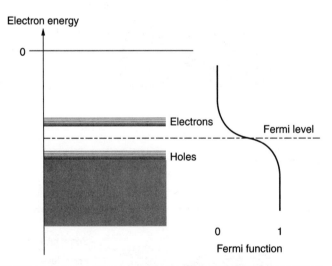

Figure 3.5   Energy band occupancy in intrinsic silicon at room temperature.

[*Note*   Again as is the case with metals, the Fermi level in an intrinsic semiconductor will change a little with temperature because of the detailed structures of the bands, but for practical purposes it can be assumed to be in the centre of the gap at all temperatures.]

3.3 VARIATION OF INTRINSIC CARRIER DENSITY WITH TEMPERATURE

The probability of occupancy of an energy level at the bottom of the conduction band can be obtained from the Fermi function by putting

$$E - E_F = E_g/2$$

giving

$$f(E) = \frac{1}{\exp(E_g/2kT) + 1}$$

Self assessment test 3.4

Estimate a value for E_g above which, assuming a temperature of 290 K, the '1' in the denominator of the above expression can be neglected.

Answer

Trying a few values of exponentials, $\exp(4) \approx 50$, $\exp(5) \approx 150$. Taking this last figure for the minimum value of the exponential function in the denominator means that the error in leaving out the 1 will be well below 1 per cent. So, $E_g/2kT$ has to be >5.

At 290 K, $2kT \approx 0.05$ eV, so E_g must be greater than 0.25 eV.

From the above you will see that in the case of silicon (and, in fact, for any semiconductor material likely to be used) I can write

$$f(E) \approx 1/\exp(E_g/2kT)$$

which I can rewrite as

$$f(E) \approx \exp(-E_g/2kT)$$

Since the probability of occupancy falls off rapidly with increasing distance from the Fermi level, most of the electrons in the conduction band will be near the bottom of the band and most of the holes in the valence band will be near its top edge. The number of conduction electrons, and holes, per unit volume will each be proportional to $f(E)$, so I can write

$$n_i = N_c\exp(-E_g/2kT)$$

N_c is a constant, sometimes called the 'density of states constant'; it can be interpreted as the equivalent number of conduction band energy states per cubic metre of the semiconductor if all the states existed at the same energy – that of the bottom of the band.

[*Note*   The expression 'density of states function' is also used in the literature to indicate the variation of the number of energy levels per unit energy interval in a band, i.e. a density in terms of energy increment rather than volume.]

Self assessment test 3.5 _____

Using values of parameters given in Table 1.2 on page 4 calculate a value of N_c for silicon.

Answer

You should get an answer $N_{c(silicon)} \approx 5 \times 10^{25}$ m^{-3}.

_____

Now I can calculate how rapidly the intrinsic carrier concentration in silicon changes with temperature around room temperature.

Taking a 1° rise in temperature,

$$n_{i(290)} = N_c\exp(-E_g/2 \times 290k), \qquad n_{i(291)} = N_c\exp(-E_g/2 \times 291k)$$

Thus

$$\frac{n_{i(291)}}{n_{i(290)}} = \frac{\exp(-E_g/582k)}{\exp(-E_g/580k)} = \exp\left[\left(\frac{E_g}{k}\right)\left(\frac{1}{580} - \frac{1}{582}\right)\right]$$

$$= 1.08$$

There is an 8 per cent rise in concentration per degree: this implies a doubling of the concentration for every 9° of temperature rise ($1.08^9 \approx 2$). It is for this reason, together with the fact that the smallest amount of impurity has a profound effect, that the value of n_i can only be quoted as an 'order of magnitude' (10^{16} m^{-3}).

3.4   CHEMICAL AND MECHANICAL DEFECTS IN SILICON

A crystal of silicon can never be entirely chemically pure. For the purpose of building semiconductor devices, it is necessary to start with material with less than one impurity atom per thousand million (1 in 10^9). The fact that this can now be produced routinely is an impressive technological achievement. The remaining impurity atoms – generally carbon, oxygen and various metals – may replace silicon atoms in the lattice or they may sit between the layers of silicon atoms as 'interstitial impurities'. In either case they affect locally the energy band structure and the mobility of carriers.

Similarly, a crystal can never be free from mechanical defects such as missing atoms or 'dislocations' (places in the crystal where a layer of atoms suddenly stops). Dislocations can be started by heat and can move through the crystal, so that the 'defect density' is in part a function of temperature. Again, dislocations affect energy band structure and mobility, so that manufacturers of device-grade silicon try to produce material with the minimum number of defects, and device manufacturers try to minimize any increase in defects caused by mechanical processing.

3.5 POLYCRYSTALLINE AND AMORPHOUS SILICON

A normally produced piece of solid silicon will not be a single crystal, but will consist of a lot of small crystals. These are still held together by covalent forces, but the structure is not continuously regular. The crystal boundaries are large defects, and the electrical properties of the solid are largely determined by this – its conductivity is unpredictable and generally much higher than that of pure single-crystal silicon at the same temperature. In certain processes, such as the deposition of silicon from a gas onto a surface under appropriate conditions, the silicon forms very small crystals. This material is called 'polycrystalline silicon' or sometimes just 'polysilicon' and is used in some devices as a conducting material.

Silicon deposited from a gas onto a surface can also, under other conditions, form an amorphous layer. Amorphous material does not have a regular crystal structure at all, and in some ways can be thought of as a very stiff liquid, the most common example in everyday life being glass. Although the atoms in amorphous silicon have variable separation, there are no sudden discontinuities as there are at the crystal boundaries of polycrystalline material, and the amorphous material can be used for certain special purposes as a rather poor semiconductor. Its only advantage is that the devices in which it can be used can be produced much more cheaply in this material than in single-crystal silicon.

I shall describe how amorphous silicon can be used to make solar cells in Chapter 12.

3.6 OTHER SEMICONDUCTOR MATERIALS

Although silicon is by far the most commonly used semiconductor material, other materials have similar properties and are used for special purposes.

The periodic table groups together elements which have the same outer electron configurations, and thus similar chemical properties. Silicon is in group 4 and has four valence electrons. The other element in group 4 with useful semiconductor properties is germanium. Group 3 elements have three valence electrons and group 5 elements have five. It is possible to make equal quantities of a group 3 and a group 5 element form a crystal together, similar in structure to a silicon crystal, in which the two types of atom in the crystal alternate and share valence electrons, averaging out at four per atom – these are known as compound semiconductors. The most useful of such combinations are gallium arsenide and indium phosphide.

Compound semiconductors will be discussed further in Chapters 10–12. Some of the properties of germanium, gallium arsenide and indium phosphide are listed in Table 1.2 on page 4.

3.7 SUMMARY

An intrinsic semiconductor solid is held together by covalent bonds. Near absolute zero of temperature there are no carriers and the material is an insulator.

At room temperature an intrinsic semiconductor material contains equal numbers of two types of current carrier: electrons, which can be modelled as negatively charged particles; and holes, which can be modelled as positively charged particles. Each of these has a measurable effective mass and mobility within the material. The conductivity of the semiconductor is given by

$$\sigma = n_i e(\mu_n + \mu_p)$$

The number of electrons and holes, n_i, is a rapidly increasing function of temperature, given by

$$n_i \approx N_c \exp(-E_g/2kT)$$

At room temperature, n_i is many orders of magnitude smaller than the carrier density in a good metallic conductor.

The allowed energies of valence and conduction electrons in a semiconductor are usefully illustrated by an energy band diagram, the significant features of which are the electron affinity, the energy band gap and the approximate position of the Fermi level at a given temperature.

The probability of occupancy of a level of energy E in the energy band diagram is given by the Fermi function

$$f(E) = \frac{1}{\exp[(E - E_f)/kT] + 1}$$

where E_F is the energy of the Fermi level.

Silicon and germanium are single-element semiconductors; compound semiconductors are also used, the most significant being gallium arsenide and indium phosphide.

3.8 PROBLEMS

Data for these problems will be found in Tables 1.1 and 1.2 on page 4.

1 Show that for each of the intrinsic semiconductor materials listed in Table 1.2, the values of the density of states constant N_c, do not differ by much more than an order of magnitude. Why would you expect this to be the case?

2 Which of the materials listed in Table 1.2 has the most rapid rise in intrinsic carrier density with temperature? Find the percentage rise in intrinsic carrier density for that material when the temperature rises by 1 °C around room temperature.

3 Gallium phosphide, another compound semiconductor, has a resistivity, for the intrinsic material, in the region of $10^{14}\,\Omega\,m$. The electron and hole mobilities in this material have approximately the same value, about $0.01\,m^2\,V^{-1}\,s^{-1}$. It may be assumed that the value of N_c is $\approx 10^{25}$.

Estimate to the nearest order of magnitude the intrinsic carrier density at room temperature in gallium phosphide, and from this deduce an approximate value for the energy band gap.

4 Find the ratio of the probability of occupancy of an energy level at the top of the conduction band (vacuum level) to that of an energy level at the bottom of the conduction band in silicon at room temperature. Hence estimate how many electrons in a cm^3 of silicon at room temperature have enough energy to escape.

4 Extrinsic semiconductors

It is possible to modify a crystal of an intrinsic semiconductor by replacing a small proportion of the original atoms in the lattice, at regular spacing, by atoms of another element with a different number of valence electrons. Such a material is then called an extrinsic semiconductor. The replacement material, usually referred to as a 'dopant', can be from group 3 of the periodic table, with three valence electrons, or from group 5, with five. Group 3 dopants are known as 'acceptor' dopants, and those from group 5 as 'donor' dopants, for reasons which will become clear.

The most commonly used acceptor dopants are boron or aluminium, the most common donors, phosphorus or arsenic.

4.1 N-TYPE SILICON

Figure 4.1 shows a two-dimensional representation of part of a crystal of silicon in which one of the silicon atoms has been replaced by a phosphorus atom. In the whole crystal, typically, about one in a million silicon atoms will have been so replaced. The phosphorus atom fits nicely, being about the same size as the silicon atoms, and supplies four valence electrons, but the fifth valence

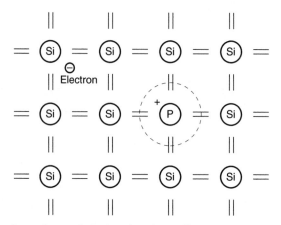

Figure 4.1 Phosphorus ion and electron in n-type silicon.

electron has no place. The phosphorus atom without the fifth electron has a net positive charge, so there is binding energy between the electron and the phosphorus atom, but this binding energy is small, 0.045 eV in fact. The phosphorus atom without the fifth valence electron can be described as a positive ion. Near absolute zero temperature the spare electron is held in the vicinity of the ion, but heat energy can easily free it, and at temperatures above about 100 K, certainly at room temperature, effectively all such electrons are freed to become conduction electrons.

If one in 10^6 of the lattice sites is occupied by a phosphorus ion, then, bearing in mind that there are 5×10^{28} lattice atoms per cubic metre, there is now a population of 5×10^{22} electrons per m³ in addition to those produced by breaking normal valence bonds. Since there are a lot more electrons per unit volume than in the intrinsic material, the chance of a hole meeting an electron and recombining is increased, so the hole population is reduced below the intrinsic value.

It is found that a simple rule applies to this case, and in fact to any case of a doped semiconductor in equilibrium. This states that:

$$n \times p = n_i^2$$

where n is the density of conduction electrons in the material and p is the density of holes.

Now, in the case that I am considering, $n = 5 \times 10^{22} + p$, since there must be as many electrons produced by broken valence bonds as there are holes. So I should write

$$(5 \times 10^{22} + p) \times p = (10^{16})^2 = 10^{32}$$

but since p must be much less than n_i, and n_i is almost seven orders of magnitude smaller than 5×10^{22}, it will make no perceptible difference to the result if I simply take n as 5×10^{22} and write

$$5 \times 10^{22} \times p = 10^{32}$$

giving

$$p = 2 \times 10^9 \, \text{m}^{-3}$$

In this material, which is called n-type, electrons are described as the 'majority carriers' and holes are called 'minority carriers'.

The conductivity is given by the equation

$$\sigma = ne\mu_n + pe\mu_p$$

When I substitute numbers for the doping described above, it is clear that the second term is negligible, and the conductivity works out to be

$$\sigma = 1200 \, \text{S m}^{-1}$$

still much smaller than that of copper, but much larger than that of intrinsic silicon.

The normal range of doping is between 1 in 10^3, which would be described as 'heavy doping', and 1 in 10^8, which would be 'light doping'. Heavily doped n-type material is often labelled n^+ in diagrams, and lightly doped material n^-; the plus and minus in this context have no other significance.

Self assessment test 4.1 _____

Calculate the carrier densities and conductivities of n-type silicon with (a) 1 in 10^3, and (b) 1 in 10^8 doping.

Solution

(a) N_D (number of donor atoms per m^3) = $5 \times 10^{28}/10^3 = 5 \times 10^{25}$.
So, $n \approx 5 \times 10^{25}\,m^{-3}$, $5 \times 10^{25} \times p \approx 10^{32}$. Therefore $p \approx 2 \times 10^6\,m^{-3}$. Now, $\sigma \approx 5 \times 10^{25} \times 1.6 \times 10^{-19} \times 0.15 = 1.2 \times 10^6\,S\,m^{-1}$.
(b) $N_D = 5 \times 10^{20}\,m^{-3}$, so $n \approx 5 \times 10^{20}\,m^{-3}$, $p \approx 10^{32}/5 \times 10^{20} = 2 \times 10^{11}\,m^{-3}$.
Thus $\sigma \approx 5 \times 10^{20} \times 1.6 \times 10^{-19} \times 0.15 = 12\,S\,m^{-1}$.

___

I need to make the point that the material as a whole remains electrically neutral. The total positive charge in the donor ions plus that in the minority carriers exactly equals that in the majority carriers.

4.2   P-TYPE SILICON

Figure 4.2 shows a two-dimensional representation of part of a crystal of silicon in which one of the silicon atoms has been replaced by a boron atom. The

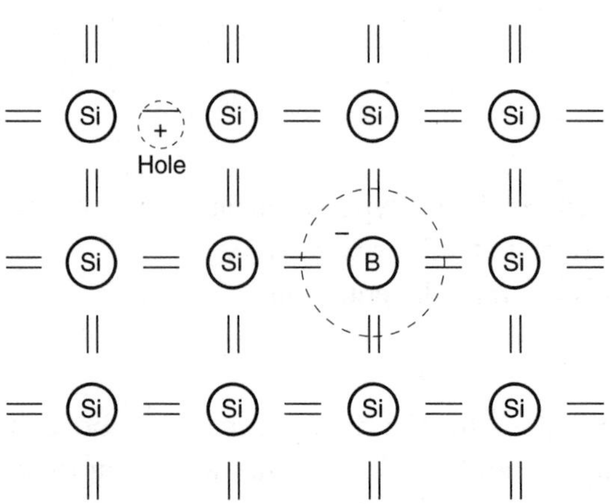

Figure 4.2   Boron ion and hole in p-type silicon.

boron atom only has three valence electrons, but a fourth one moves in, making the boron a negative ion, and a hole is generated. The hole is weakly bound electrostatically to the boron ion – again, as it happens, with an energy of 0.045 eV – and breaks free at well below room temperature.

The properties of p-type silicon are similar to those of n-type silicon with the same doping density, the main differences being the signs of the majority and minority carriers and the values of conductivity, which are lower because of the lower mobility of holes. The boron atoms are called acceptors because they 'accept' (as opposed to donate) a valence electron. The symbol normally used for the acceptor ion density is N_A m^{-3}.

Again, p-type silicon is electrically neutral, containing negative charge in the acceptor ions and in the minority carriers, balancing the positive charge in the majority carriers.

4.3 DOPED SEMICONDUCTORS AT HIGH TEMPERATURE

As the temperature is increased, the number of carriers created by broken valence bonds increases rapidly, until finally the number of holes and electrons in the material is almost equal. The excess carriers are swamped and the material is said to have reverted to intrinsic behaviour. The limiting temperature for extrinsic behaviour is usually taken as that temperature at which the intrinsic carrier density in the pure material would be the same as the majority carrier density at room temperature (i.e. as the doping density); this is quite arbitrary, but it gives some sort of reference temperature. I shall consider the most lightly doped material, where the minority carrier density is highest, and so, taking n-type silicon with 5×10^{20} donor atoms per m^3 (i.e. with 1 in 10^8 doping) as an example, at the limiting temperature,

$n_i = 5 \times 10^{20}$ m^{-3}

Now, $n_i = N_c \exp(-E_g/2kT)$, and I showed in Chapter 3 that for silicon, $N_c \approx 5 \times 10^{25}$ m^{-3}, so I need to solve

$5 \times 10^{20} = 5 \times 10^{25} \exp[-1.12/(2 \times 86.2 \times 10^{-6} \times T_L)]$

This can be rearranged as

$\exp(6497/T_L) = 10^5$

Taking natural logarithms,

$6497/T_L = 11.5$

$$T_L \approx 565\text{K} \approx 290\,°\text{C}$$

Self assessment test 4.2

Work out roughly the ratio of majority to minority carrier densities in the material considered above at a temperature of 200 °C.

Answer

n_i works out to approximately $5 \times 10^{19}\,\mathrm{m^{-3}}$. Considering as an approximation $5 \times 10^{20} \times p = (5 \times 10^{19})^2$ gives $p \approx 5 \times 10^{18}$.

Checking back, you will find that even at this level of n_i there isn't much error in the approximation, so $n/p \approx 100$.

In fact a temperature of 200 °C represents a reasonable upper limit at which to use devices made of silicon.

4.4   ENERGY BAND DIAGRAMS FOR EXTRINSIC SEMICONDUCTORS

Figure 4.3 shows the energy band diagram for n-type silicon. Compared to the diagram for intrinsic silicon there is an extra level, in the gap near to the bottom of the conduction band, representing the energies of bound donor electrons. If the dopant is phosphorus, its distance below the conduction band edge is 0.045 eV. Because this level only occurs in the vicinity of a donor ion it is shown as a dashed line. This also carries the implication that, from the point of view of quantum mechanics, the donor ions are sufficiently remote from each other that each bound donor electron can be at effectively the same level and each line segment can accommodate one electron. The horizontal axis on the diagram has now acquired a meaning – it indicates distance through the crystal in a chosen direction.

At above a temperature of 100 K or so, virtually all the donor levels are

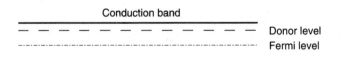

Conduction band

Donor level

Fermi level

Energy gap

Valence band

Figure 4.3   Donor level and Fermi level in an n-type semiconductor at room temperature.

empty, their electrons having been promoted by heat to the conduction band. At higher temperatures a few of the donor levels may be refilled from the valence band; the total distribution of electron energies can be deduced from the Fermi function. The Fermi level moves significantly within the energy gap as the temperature changes: at low temperature it lies between the conduction band edge and the donor level; at room temperatures it is just below the donor level, consistent with the fact that a significant number of valence band electrons are jumping to the conduction band; at high temperatures, as the material approaches the intrinsic condition, the Fermi level moves towards the centre of the gap. In Fig. 4.3, the Fermi level is shown in the sort of position that it will be in at room temperature.

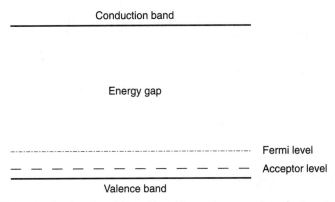

Figure 4.4   The acceptor level and Fermi level in a p-type semiconductor at room temperature.

Figure 4.4 shows the band diagram for p-type silicon. It is the complement of the n-type silicon diagram. In this case the acceptor level is near the valence band edge (0.045 eV above if the dopant is boron) and is effectively full above 100 K with electrons which have moved up from the valence band, leaving holes. The Fermi level moves as temperature rises from a position between the valence band edge and the acceptor level towards the middle of the gap.

Donor levels and acceptor levels are frequently described in the literature as 'impurity' levels. Since the dopants are deliberately added, you might find it misleading to describe them as impurities. It is the case, however, that all foreign atoms, whether added deliberately or not, introduce localized energy levels in the gap.

There is no contradiction between the energy band description and the dynamic description of the behaviour of electrons and holes in extrinsic semiconductors. They simply give different perspectives on the phenomena and are useful in explaining different facets of the material's properties.

4.5   COMPENSATORY DOPING

Suppose that both donor and acceptor impurities are added to an intrinsic semiconductor. If they are added in equal amounts then they will cancel each other's effect, and the material will behave as if it were intrinsic. From the point of view of the energy band diagram there will be both donor and acceptor levels in the gap, but since the empty acceptor levels are at lower energy than the full donor levels, electrons will fill the acceptor levels by falling from the donor levels rather than by promotion from the valence band. If $N_A > N_D$ then the material will behave as p-type with an effective doping density of $N_A - N_D$; similarly, if $N_D > N_A$ the material will be n-type with effective doping density $N_D - N_A$.

The fact that doping can be compensated in this way is very useful when semiconductor devices requiring different doping in different parts need to be made. The only limiting factor is that the total density of dopant atoms must not be too high; if it is, then carrier mobilities will be decreased, the dislocation density will be greater and the crystal structure may be distorted (and with this the energy band structure) to such an extent that the material may behave more like a metallic conductor than a semiconductor.

4.6   GRADED DOPING

It often occurs in practice that a semiconductor is doped in such a way that the density of dopant ions varies from one part of the crystal to another. Under these circumstances a permanent electric field arises in the crystal and there are voltage differences between different parts of the crystal. I shall first explain the effect qualitatively and then give a mathematical analysis.

Figure 4.5 shows a crystal of p-type silicon in which the acceptor density varies smoothly from 10^{22} m^{-3} at one end to 10^{25} m^{-3} at the other. I shall take the length of the crystal to be a few μm, which is typical of the distances involved in silicon integrated circuit technology. If one imagines the moment at which the material is formed, all parts of the crystal are electrically neutral, but

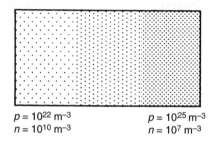

$p = 10^{22}$ m^{-3} $p = 10^{25}$ m^{-3}
$n = 10^{10}$ m^{-3} $n = 10^{7}$ m^{-3}

Figure 4.5   A p-type silicon crystal with N_A varying from 10^{22} at one end to 10^{25} at the other.

the initial density of holes varies from 10^{22} to 10^{25} m^{-3} along the length of the piece, while that of electrons varies from 10^{10} to 10^7 m^{-3} in the same direction. This is not a stable condition; the carriers, being mobile, will diffuse away from the regions in which they have higher density.

The phenomenon of diffusion occurs in all populations of mobile particles. It is a random process, driven by heat, and is not caused by any force. If, for example, a box with two interconnectable chambers, such as that shown in Fig. 4.6, contains oxygen in one and nitrogen in the other at the same pressure, then if the door separating the two is opened, in a very short time both chambers will contain a mixture of oxygen and nitrogen. All the molecules are moving around energetically, and so long as there are more oxygen molecules in the left-hand chamber than in the right, more oxygen molecules per second in the left-hand chamber will by chance arrive at the opening and pass through than will do so in the right-hand chamber. Equilibrium will be achieved when the density of oxygen molecules is the same on both sides. A similar argument holds for the nitrogen molecules.

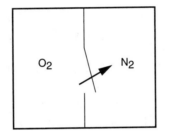

Figure 4.6

In the case of the semiconductor that I am discussing, the carriers will not achieve a uniform distribution because they carry charge. Holes migrating to the lightly doped end make that end positively charged and leave the other end negatively charged. The electrons, in migrating the other way, add to the effect by carrying negative charge to the heavily doped end. The diffusion must stop at the surface, so surface layers of charge are built up which create an electric field inside the material. This drives the carriers the other way, so that an equilibrium is reached in which the tendency for carriers to diffuse is exactly balanced by the tendency of carriers to drift in the internal field.

Since there is an electric field inside the material, there must be voltage differences between different points in the material. In proceeding to find the value of this voltage I shall interpret the equilibrium condition as equal and opposite currents of electrons and equal and opposite currents of holes – with a resultant current of zero.

4.7 DIFFUSION CURRENT

In Chapter 1, I described the mechanism of drift current, that is the electric current caused by an electric field; this is what one thinks of as the normal process of conduction. There is, however, as is implied in the previous section, another process which will cause an electric current to flow in which the driving mechanism is diffusion.

The significant factor which determines diffusion current is the density gradient. Theoretical analysis of the process of diffusion shows that, where the variation in particle density is in one direction only, the rate of flow of diffusing particles is proportional to the rate of change with distance of the particle density. In symbols this is represented as

$$\Phi = -D\frac{dN}{dx}$$

where Φ is the number of particles per second crossing unit area normal to the direction of the density gradient, N is the number of particles per unit volume at a given point and x is the distance along the direction of density gradient. D is called the diffusion constant (don't confuse it with electric flux density). The minus sign is there because by convention dN/dx indicates the *increase* in density with distance, but particles diffuse *down* the density gradient.

If the particles are electrons or holes, then $e\,\Phi$ is a current density, so in that case I can write, using the usual symbols and subscripts,

$$J_{n(dif)} = eD_n\frac{dn}{dx}$$

for electrons, and

$$J_{p(dif)} = -eD_p\frac{dp}{dx}$$

for holes. In the case of the electrons, the fact that conventional current is in the opposite direction to particle flow eliminates the minus sign from the formula.

It is reasonable to assume that the mechanisms which resist the movements of electrons or holes which are diffusing will be the same as those which impede the flow of drifting electrons of holes, so one would expect to find a relationship between the diffusion constant D and the mobility μ for an electron or hole. This relationship proves to be

$$D = \frac{kT}{e}\mu$$

and is known as Einstein's relation, since it was he who derived it.

Self assessment test 4.3_____

Calculate the value of kT/e at room temperature, and confirm that the values given for carrier mobilities and diffusion constants in Table 1.1, page 4, are consistent.

Answer

$kT/e = 0.025$ V. You should have found that the given values are consistent, allowing for the fact that rounding errors will have been introduced by expressing the values to only two significant figures in the table.

_____

As I showed in Chapter 1 that the flow of current in a copper wire with an applied emf implies some variation with distance of electron density, some of the current must be diffusion current. The amount of diffusion current, however, is very small, and the effect is hidden. We shall see that in considering the flow of current in extrinsic semiconductors, diffusion current is often very important.

4.8 VOLTAGE DIFFERENCE BETWEEN POINTS IN AN EXTRINSIC SEMICONDUCTOR WITH DIFFERENT DOPING DENSITIES

In the following derivation I am going to have to be very careful about plus and minus signs. I am going to start by considering holes, because they flow in the same direction as conventional current, which eliminates one source of confusion.

Figure 4.7 represents a piece of silicon with p-type doping which varies from light at the left-hand end to heavy at the right-hand end. I am going to consider two points in the crystal, a and b as shown. Diffusion has created an electric field in the direction from a to b (the positive x direction) and the density

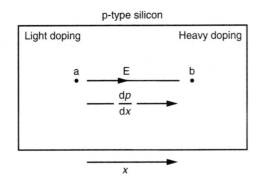

Figure 4.7

gradient of holes is positive in that direction, since the number of holes increases as we go from left to right.

In the equilibrium condition the total current due to holes at every point in the crystal is zero.

Let the electric field at a point x be E and the hole density at that point be p. The drift current density of holes in the electric field is

$$e\mu_p p E$$

The diffusion current density at the same point is

$$-eD_p \frac{dp}{dx}$$

(the minus shows that the diffusion current is going the other way). The sum of these currents is zero, so that

$$e\mu_p p E - eD_p \frac{dp}{dx} = 0$$

also

$$Dp = \frac{kT\mu_p}{e}$$

together giving, after cancelling terms on both sides,

$$epE = kT\frac{dp}{dx}$$

Now

$$E = \frac{-dV}{dx}$$

(the minus sign shows that the electric field is in the direction in which voltage is falling). Thus, with a little rearrangement,

$$-\frac{e}{kT}\frac{dV}{dx} = \frac{1}{p}\frac{dp}{dx}$$

If I integrate this equation between the points a and b, I can call the limits V_a and V_b for voltage and p_a and p_b for hole density:

$$-\int_{V_a}^{V_b} \frac{e}{kT}dV = \int_{p_a}^{p_b} \frac{1}{p}dp$$

The result of this integration is

$$-\frac{e}{kT}(V_b - V_a) = \ln\left(\frac{p_b}{p_a}\right)$$

which can be rearranged as

$$V_a - V_b = (kT/e)\ln(p_b/p_a)$$

The voltage difference between two points, a and b, can be expressed conventionally as V_{ab}, meaning the voltage of a with respect to b, so I can rewrite the equation as

$$V_{ab} = (kT/e)\ln(p_b/p_a) \tag{4.1}$$

By a similar argument (but you have to be even more careful with pluses and minuses), considering electron drift and diffusion currents it can be shown that

$$V_a - V_b = (kT/e)\ln(n_a/n_b)$$

that is

$$V_{ab} = (kT/e)\ln(n_a/n_b) \tag{4.2}$$

Notice that no assumptions are made about how the electric field or carrier densities vary with distance. The results are absolutely general and apply to a variation of carrier densities caused by any sort of doping variation, even to a change from acceptor to donor doping, in the same crystal; this latter condition will be considered in the next chapter.

Taking Equations (4.1) and (4.2) together, I can deduce that

$$\frac{p_b}{p_a} = \frac{n_a}{n_b}$$

so

$$p_a n_a = p_b n_b$$

Since a and b can be anywhere in the crystal, and the doping can vary in any way, including zero doping (in which case p and n are both n_i), this confirms the earlier quoted result, that anywhere in a semiconductor in equilibrium

$$pn = n_i^2$$

Returning to Fig. 4.5, I can now find the voltage between the ends of the specimen. If the doping gradient is uniform, so that the initial density gradients are uniform, then the excess charges will be at the end surfaces and will not be sufficiently large significantly to affect the majority carrier densities. Hence, in Equation (4.1), I can assume $p_a = N_{Aa}$ and $p_b = N_{Ab}$, where N_{Aa} and N_{Ab} are the acceptor densities at the two ends.

Since $N_{Aa} = 10^{22}$ atoms m^{-3} and $N_{Ab} = 10^{25}$ atoms m^{-3}, I get from Equation (4.1)

$$V_{ab} = 0.025 \times \ln(10^3) = 0.17\ \text{V}$$

4.9 SUMMARY

Extrinsic semiconductors at room temperature have more of one type of carriers than of the other: n-type has a majority of electrons and a minority of holes; p-type has a majority of holes and a minority of electrons.

If the temperature is raised sufficiently, the numbers of carriers become very much greater and more equal – the material approaches the 'intrinsic condition'.

n-type material is made by doping intrinsic material with donor atoms, which have five valence electrons; p-type material is made by doping intrinsic material with acceptor atoms, which have three valence electrons. If both types of impurity are present, the material behaves as n or p type according to which impurity is in the greater concentration.

At normal working temperatures, for n-type material,

$$n \approx N_D$$

and for p-type material,

$$p \approx N_A$$

For both cases, everywhere in the crystal

$$n \times p = n_i^2$$

provided the material is in equilibrium – what this proviso implies will become clear in subsequent chapters.

The donor atoms are ionized, by losing or gaining an electron respectively, but overall the material is electrically neutral, i.e. the sum of minority carriers and ions equals the number of majority carriers.

If the doping is not uniform throughout the crystal, then, notwithstanding the overall charge neutrality, there are electric fields in the material as a result of the diffusion of carriers. The voltage of a point a in the crystal with respect to a point b is related to the carrier densities by the equations

$$V_{ab} = (kT/e) \ln(p_b/p_a)$$

and

$$V_{ab} = (kT/e) \ln(n_a/n_b)$$

When energy band diagrams are drawn for extrinsic semiconductors the horizontal axis is used to indicate distance in one dimension (in a direction chosen because it has some sort of significance) through the crystal. Compared to energy band diagrams for intrinsic materials, those for extrinsic máterials have extra levels, appearing in the energy gap in the vicinity of dopant atoms.

The impurity level produced by n-type doping is very close to the conduction band edge and is effectively empty at room temperature, having given up its electrons as majority carriers. The impurity level produced by p-type doping is very close to the valence band edge and is effectively full at room temperature with electrons which have left the valence band leaving behind holes as majority carriers.

The Fermi level in extrinsic material moves from a position between the impurity level and the band edge, which it occupies at very low temperatures, to the middle of the band at very high temperatures. At room temperature it is usually near to the impurity level on the mid-band side.

4.10 PROBLEMS

1 A piece of pure germanium, of volume 1 mm^3, is uniformly doped with aluminium to a concentration of 1 in 10^5. The packing density of atoms in crystalline germanium is $\approx 4 \times 10^{28} \text{ m}^{-3}$.
   (a) Estimate how many aluminium atoms are involved.
   (b) Estimate how many majority carriers and how many minority carriers there are in the specimen at room temperature.
   (c) Estimate the limiting temperature for extrinsic behaviour for this specimen, and indicate how many of each type of carrier it will contain at that temperature.

2 A conductive strip in an integrated circuit is to be formed from a layer of single-crystal silicon $5 \ \mu\text{m}$ thick, the strip being $15 \ \mu\text{m}$ wide and $100 \ \mu\text{m}$ long. The dopant to be used is to be phosphorus. Calculate the doping density in parts per million required if the resistance of the strip is to be no greater than $10 \ \Omega$.

3 A layer of silicon in an integrated circuit is doped with arsenic by a process which results in the doping density on one face of the layer being 10^{25} atoms m^{-3}, reducing to 10^{20} m^{-3} at the other face. Deduce the direction of the internal electric field and estimate the value of the potential difference between the faces.

5 PN junctions

A pn junction occurs in a single crystal of semiconductor in which there is a transition from p-type material to n-type. For the purposes of analysis in this chapter I am going to assume that the transition is abrupt with uniform doping on either side, although this is by no means always the case in practice.

The explanations and derivations which follow are what might be described as 'first order', by which I mean that I have tried to simplify as much as possible so as to give a correct impression of properties and functions, while ignoring all sorts of 'ifs and buts'. Later I shall point out, as appropriate, where and why actual device behaviour deviates from this (relatively) simple theory.

For many of the worked examples in this chapter I use the same junction which I shall define here so that I can refer back to it:

Example junction: an abrupt pn junction in silicon between doping with 10 parts per million of boron and doping with 1 part per million of phosphorus.

Some of the following arguments should now be familiar, since they are similar to those used in the last sections of Chapter 4, but I shall repeat them for clarity.

5.1 AN ABRUPT PN JUNCTION IN SILICON

Figure 5.1 represents a crystal of silicon in which one half is doped with boron and the other with phosphorus. You have to imagine that this is the instant at which the crystal is formed, so that on the p side there is a uniform distribution of many holes and a few electrons, while on the n side the situation is reversed. This is an unstable condition: holes will diffuse to the n side and electrons to the p side. The transport of charge associated with this movement of carriers results in the n side becoming positively charged and the p side negatively charged. From this there arises an electric field which prevents further diffusion.

How must the charges be distributed when the crystal is in equilibrium? Following from electrostatic considerations, as described in Chapter 2, when two oppositely charged regions are in close proximity the excess charges must reside as near as possible to the boundary. So, in this case, negative charge in

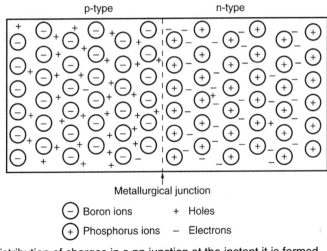

p-type   n-type

Metallurgical junction

⊖ Boron ions   +   Holes
⊕ Phosphorus ions   −   Electrons

Figure 5.1   Distribution of charges in a pn junction at the instant it is formed.

the p material and positive charge in the n material must be as near as possible to the metallurgical junction. However, apart from a very few electrons, the only negative charges in the p material are the donor ions, and similarly the positive charges in the n-type material are mostly acceptor ions. Hence, on each side of the junction, the carriers move back to expose the ions in layers which together are known as the transition region (also sometimes called the depletion layer). The resultant electric flux density due to these charges exists only in the transition region, in the same way as it only exists between the plates of a charged capacitor, because it cancels elsewhere, so that the electric field which holds back diffusion is purely in the transition region. The regions on either side of the transition region have the normal populations of majority and minority carriers for the material and temperature, and are known as the neutral regions. The distribution of charges is shown diagramatically in Fig. 5.2.

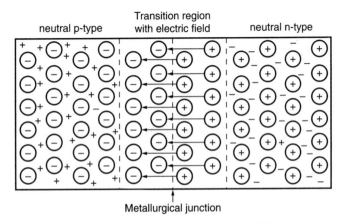

                    Transition region
neutral p-type   with electric field   neutral n-type

Metallurgical junction

Figure 5.2   Electric charges and fields in a pn junction in equilibrium.

The situation described is very much a dynamic equilibrium. Carrier pairs are constantly being created by heat and recombining. Any minority carriers which happen to enter the transition region are swept across to the other side by the field which is in the right direction to accelerate them. To balance this, majority carriers with higher than average energy can overcome the potential barrier (i.e. cross against the electric field). Clearly, the idea that there are no carriers in the transition region is an oversimplification.

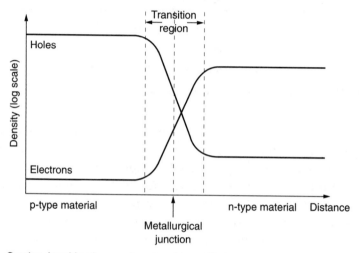

Figure 5.3   Carrier densities in a pn junction in equilibrium.

Figure 5.3 shows approximately how the carrier densities change through the transition region. The junction considered here has a more heavily doped p region than n. Notice that the scale is logarithmic, so the falls in density are more rapid than they appear. For many purposes it is a valid approximation to assume that the transition region has definite boundaries and contains virtually no carriers.

5.2   THE CONTACT POTENTIAL, ψ

The contact potential or barrier potential is the voltage difference between the two sides of the junction associated with the electric field in the transition region. This can be found using equations, developed in Chapter 4, which apply to any single crystal of semiconductor material:

$$V_{ab} = (kT/e) \ln(p_b/p_a) \tag{5.1}$$

or

$$V_{ab} = (kT/e) \ln(n_a/n_b) \tag{5.2}$$

For pn junctions, by convention, voltages are always measured from the p side to the n side. Since each neutral region is at a uniform potential, any point in the p neutral region will do for the point a, and any point in the n neutral region will do for b.

There is also a useful convention for designating carrier densities: a first subscript is used for the region and a second subscript indicates the state – '0' indicating equilibrium. Hence

- p_{p0} = the density of holes (majority carriers) in the p region in the equilibrium condition;
- p_{n0} = the density of holes (minority carriers) in the n region in the equilibrium condition;
- n_{n0} = the density of electrons (majority carriers) in the n region in the equilibrium condition;
- n_{p0} = the density of electrons (minority carriers) in the p region in the equilibrium condition.

When later we meet p_{n1}, for example, this will represent the density of holes in the n-type region in non-equilibrium conditions at a point designated '1' in the material.

Applying Equation (5.1), then, I can write

$$\psi = (kT/e) \ln(p_{n0}/p_{p0}) \tag{5.3}$$

and from Equation (5.2)

$$\psi = (kT/e) \ln(n_{p0}/n_{n0}) \tag{5.4}$$

Now p_{p0} and n_{n0} are effectively equal to N_A and N_D, the acceptor density in the p material and the donor density in the n material, respectively, and n_{p0} and p_{n0} can each be replaced by n_i^2/N_A and n_i^2/N_D, respectively. So, from Equation (5.3),

$$\psi = kT/e) \ln[n_i^2/(N_A N_D)] \tag{5.5}$$

Substituting in Equation (5.4) gives the same result.

Self assessment test 5.1

For the junction described as 'example junction' at the beginning of the chapter, find the contact potential at room temperature.

Solution

$N_A = 5 \times 10^{28}/10^5 = 5 \times 10^{23}$ m^{-3}. Similarly $N_D = 5 \times 10^{22}$ m^{-3}, $kT/e = 0.025$ V, $n_i^2 = 10^{32}$. Thus

$\psi = 0.025 \ln[10^{32}/(5 \times 10^{23} \times 5 \times 10^{22})]$

$\psi = -0.83$ V

You will have found that ψ is negative. In equilibrium the p side is negative to the n side and so the contact potential must be negative because of the convention adopted for the direction of voltage across a pn junction.

5.3   THE ENERGY BAND DIAGRAM FOR A PN JUNCTION

Figure 5.4 shows the significant parts of the energy band diagram for a pn junction (I have not bothered to show the bottom of the valence band). Equilibrium requires the Fermi level to be the same throughout the crystal. The net negative charge on the p side lifts all the energy levels of the electrons on the p side, and the net positive charge on the n side results in the energy levels on the n side being lowered. Notice that the band edges are bent through the transition region.

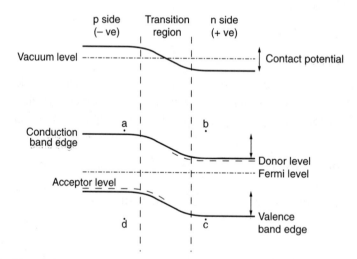

Figure 5.4   Energy band diagram for pn junction in equilibrium (the bottom of the valence band is not shown).

[**Note**   You should be able to see from this why two dissimilar metals in contact may produce a contact potential. If their work functions – the energy difference between the Fermi level and the vacuum level – are different, they will exchange charge on contact to bring their Fermi levels in line. There will not be a significant transition region, simply a double layer of charge at the contact.]

Several useful ideas can be gleaned from Fig. 5.4. You should be able to see that the size of the contact potential will normally be less than the band gap E_g. As the temperature rises, since the Fermi level moves nearer to the middle of the gap on each side, the value of ψ should be smaller: inspection of Equation (5.5) seems to contradict this, but remembering that n_i changes rapidly with T, it is the case.

Self assessment test 5.2

For the example junction, estimate the contact potential at a temperature of 400 K.

Solution

At 400 K,

$$ni = N_c \exp(-E_g/2kT)$$

$$\approx 5 \times 10^{25} \exp[-1.12/(2 \times 86.2 \times 10^{-6} \times 400)]$$

$$\approx 4 \times 10^{18} \text{ m}^{-3}$$

$$kT/e = 8.62 \times 10^{-5} \times 400 \text{ V}$$

$$\approx 0.034 \text{ V}$$

$$\psi \approx 0.034 \ln[16 \times 10^{36}/(5 \times 10^{23} \times 5 \times 10^{22})]$$

$$\psi \approx -0.72 \text{ V}$$

In the last example we calculated ψ for this junction at 290 K. You will see that an increase in temperature has reduced its value.

Returning to Fig. 5.4, an energy level on the p side, indicated in position and energy by the point a, has the same probability of occupancy as one on the n side indicated by the point b, since they are both the same distance above the Fermi level. An electron at a and an electron at b are not, however, in the same condition. The electron at a is a minority carrier and, since it is near the bottom of the conduction band, has a low kinetic energy. If the electron at point 'a' should move towards the transition region without making any collisions it will be swept into the n region by the electric field, and if it continues to make no collisions it can reach b. Here it has more kinetic energy, as indicated by its being higher up the conduction band, but less potential energy, so that its total energy is unchanged: the transfer of energy from potential to kinetic has been brought about by the electric field. An electron at b is an unusually energetic majority carrier and thus would be capable, if moving in the correct direction, of surmounting the potential barrier by 'climbing' the electric field and reaching the point a where it would then be a minority carrier with less kinetic energy but more potential energy. The dynamic equilibrium involves constant interchanges of this sort.

A similar, but inverse, story relates to holes. A hole, being a positive particle, has greater energy the lower it is in the band diagram (nearer to the positive core atoms, which can be perceived as repelling it). Again, distance from the band edge – in this case the top edge of the valence band – can be related to kinetic energy. A hole at the point c is a minority carrier with low kinetic energy. It could move to position d, in which case it would become a majority carrier with greater kinetic energy and less potential energy. Conversely, a hole at d, in moving to c, gives up kinetic energy in the electric field but gains potential energy.

If an external voltage (from, say, a battery) were connected across the crystal with the positive pole to the p side and the negative to the n side, this would make the p side less negative and the n side less positive and so push down the energy levels on the p side and raise the energy levels on the n side. This would lower the potential barrier, and the probability of electrons moving from n to p would be raised, while that of electrons moving from p to n would be reduced – the equilibrium would be disturbed and a net current would flow. Applying the voltage the other way round would also disturb the equilibrium, and again a current would flow. However, in the first case there are many electrons on the n side ready to cross to the p side and many holes on the p side ready to flow to the n side, whereas in the second case electrons on the p side and holes on the n side are few, so the two currents are not the same. With a 'forward' voltage, i.e. positive to p and negative to n, a large current flows, whereas with 'reverse' voltage only a small current flows.

The property of allowing more current to flow with a voltage applied in one direction than if it were applied in the reverse direction is known as 'rectification' and is the most important property of a pn junction. Although I can use the energy band diagram to explain the phenomenon, I cannot get from it a formula which will relate the magnitude of current to the applied voltage; for that I need to go back to considering carrier dynamics.

5.4   THE STEADY-STATE CURRENT/VOLTAGE RELATIONSHIP FOR A PN JUNCTION

The following analysis depends critically on assumptions about how long different things take to happen. I also have to adopt various simplifications – otherwise the analysis would be impossibly complex – and I shall try to justify these as I go along.

Consider a pn junction in a crystal to which a forward voltage V_D is applied. I shall deal with the holes first. Look at Fig. 5.5: the holes in the p region next

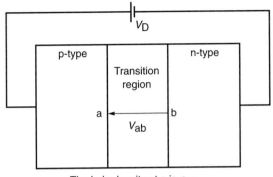

The hole density at a is p_{p0}
The hole density at b is p_{n1}

Figure 5.5

to the transition layer will quickly diffuse across to the n side so that an excess of holes is established next to the transition region edge on the n side. The sudden appearance of a lot of positive charge on the n side will tend to set up large electric fields, but these will quickly be neutralized by majority carriers moving to cancel the excess charge – not by recombination, but by their presence. This rearrangement of charges in a conductor in response to an attempt to set up a field is known as dielectric relaxation and takes place in a very short time. The average length of time that an individual hole exists before it recombines with an electron is known in this context as the minority carrier lifetime, and has a value, typically, of the order of 10^{-7} seconds: this is many orders of magnitude greater than the dielectric relaxation time.

Now the hole density in the n material next to the transition region edge has increased by a significant amount compared to p_{n0}, so I shall call the new value p_{n1}. The hole density in the p material next to the transition region has also risen, by the same process of dielectric relaxation as described above, to neutralize the field of extra electrons that have diffused to the p side, but this change is small compared to the majority carrier density so I shall ignore it and assume that the hole density on the p side is still p_{p0}.

Assuming that a new equilibrium is established in which a new diffusion current across the transition region balances a new drift current due to the changed electric field I can assume that Equation (5.1) can again be applied, giving

$$V_{ab} = \frac{kT}{e} \ln\left(\frac{p_{n1}}{p_{p0}}\right)$$

Now,

$$V_{ab} = V_D + \psi$$

so

$$\ln\left(\frac{p_{n1}}{p_{p0}}\right) = \frac{eV_D}{kT} + \frac{e\psi}{kT}$$

But, from Equation (5.3),

$$\frac{e\psi}{kT} = \ln\left(\frac{p_{n0}}{p_{p0}}\right)$$

thus

$$\frac{eV_D}{kT} = \ln\left(\frac{p_{n1}}{p_{p0}}\right) - \ln\left(\frac{p_{n0}}{p_{p0}}\right)$$

which, using the properties of logarithms, gives

$$\frac{eV_D}{kT} = \ln\left(\frac{p_{n1}}{p_{n0}}\right)$$

This can be rearranged to give

$$p_{n1} = p_{n0}\exp(eV_D/kT) \tag{5.6}$$

Exactly similar reasoning for electrons leads to

$$n_{p1} = n_{p0}\exp(eV_D/kT) \tag{5.7}$$

These two equations indicate how a forward bias changes the minority carrier densities at the transition region edges. The two equations can also be made to relate to reverse bias by giving V_D a negative value.

Self assessment test 5.3

For example junction, at room temperature, calculate the minority carrier density at the depletion layer edge on the n side (a) for a forward bias of 0.5 V, (b) for a reverse bias of 0.5 V.

Solution

(a) At room temperature, $e/kT = 1/0.025 = 40 \text{ V}^{-1}$. Now

$$p_{n0} = n_i^2/N_D = 10^{32}/5 \times 10^{22} = 2 \times 10^9 \text{ m}^{-3}$$

Using Equation (5.6),

$$p_{n1} = 2 \times 10^9 \times \exp(40 \times 0.5) \approx 10^{18} \text{ m}^{-3}$$

(b) Again using Equation (5.6), but putting $V_D = -0.5$ V,

$$p_{n1} = 2 \times 10^9 \times \exp(40 \times -0.5) \approx 4 \text{ m}^{-3}, \quad \text{i.e. effectively zero}$$

The above calculation will illustrate that for anything more than quite a small reverse bias the minority carrier densities at the depletion layer edges are effectively zero.

Returning to the forward bias case, what will happen to the excess minority carriers that have been transferred across the transition region? They will diffuse away from the junction and ultimately recombine with majority carriers. As they diffuse away the equilibrium requires them to be replaced, so there is a steady current across the transition region and through the crystal. There seems to be a contradiction here, since in describing the new equilibrium I assumed that there is no resultant current across the transition region. In fact, the balance between the drift and diffusion currents in the transition region is between two relatively large, almost equal and opposite, currents, and the slight excess of diffusion which results in the net current in the junction does not materially alter the results that I derived.

I still have not arrived at a formula for the current in the junction; to do so I need to make some assumptions about the widths of the neutral regions and about how exactly the voltage has been applied to the neutral regions. When minority carriers diffuse in an extrinsic semiconductor, the average distance that they travel before recombining with majority carriers is called the *diffusion length*, symbol L_n or L_p depending on the carrier type. The diffusion length depends on the diffusion constant and on the minority carrier lifetime; in fact it can be shown that

$$L_n = \sqrt{(D_n t_n)} \quad \text{and} \quad L_p = \sqrt{(D_p t_p)}$$

for electrons and holes, respectively. If the distance across a neutral region in the crystal is shorter than the diffusion length for minority carriers in that region, then it can be assumed that little recombination occurs in the neutral region and this leads to the result that in the steady state the diffusion gradient in the neutral region is constant.

Self assessment test 5.4

Assuming a minority carrier lifetime of the order of 10^{-7} seconds in each case, estimate the diffusion length of electrons and of holes as minority carriers in silicon.

Answer

For electrons, $L_n \approx \sqrt{(0.0039 \times 10^{-7})} = 20\,\mu\text{m}$.
For holes, $L_p \approx \sqrt{(0.0012 \times 10^{-7})} = 11\,\mu\text{m}$.

The solution to the above suggests that for the assumption of little recombination to be valid, dimensions have to be of the order of $10\,\mu\text{m}$ or less.

The other significant question is what happens to the carriers when they reach the edge of the crystal? To apply the voltage, there has to be a contact to the crystal, normally a metal one, and at such a contact the distortion of the crystal structure causes carriers to have very short lifetimes. The result of this is that the carrier densities can never depart significantly from the equilibrium value: excess minority carriers that arrive there rapidly recombine with majority carriers. Such a contact is sometimes referred to as a 'surface of infinite recombination'.

I am now in a position to work out the diffusion currents in a pn junction with short neutral regions so that there is negligible recombination until the minority carriers reach the end contacts. I shall also assume that the shape of the crystal and the areas of the contacts are such that the carriers move in straight lines perpendicular to the junction and thus the current density is constant. Starting with the holes in the n-type neutral region, I can draw a minority carrier profile as shown in Fig. 5.6. To justify the assumption that the steady-state density gradient must be constant you simply need to note that the current consists entirely of diffusing holes, which are constant in number since they do not recombine. Thus for continuity of current density, from

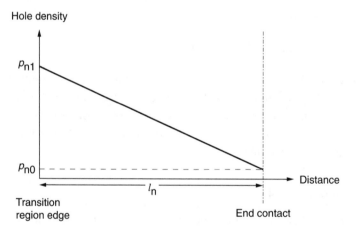

Figure 5.6

$$j_{\text{dif}} = eD_{\text{p}}\frac{\text{d}p}{\text{d}x}$$

$\text{d}p/\text{d}x$ must be constant. From Fig. 5.6,

$$\frac{\text{d}p}{\text{d}x} = \frac{p_{\text{n1}} - p_{\text{n0}}}{l_{\text{n}}}$$

where l_{n} is the length of the n-type neutral region. Also, from Equation (5.6)

$$p_{n1} = p_{n0} \exp\left(\frac{eV_D}{kT}\right)$$

So the hole current is

$$I_p = \frac{AeD_p p_{n0}}{l_n}\left[\exp\left(\frac{eV_D}{kT}\right) - 1\right]$$

By an exactly equivalent argument, the electron current in the p region is given by

$$I_n = \frac{AeD_n n_{p0}}{l_p}\left[\exp\left(\frac{eV_D}{kT}\right) - 1\right]$$

These two currents add to give the total current across the junction, giving a total, usually given the symbol I_D:

$$I_D = Ae\left[\frac{D_p p_{n0}}{l_n} + \frac{D_n n_{p0}}{l_p}\right]\left[\exp\left(\frac{eV_D}{kT}\right) - 1\right]$$

I am now going to define a current, I_s, and write

$$I_s = Ae\left[\frac{D_p p_{n0}}{l_n} + \frac{D_n n_{p0}}{l_p}\right] \tag{5.8}$$

so that

$$I_D = I_s\left[\exp\left(\frac{eV_D}{kT}\right) - 1\right] \tag{5.9}$$

Suppose V_D has a negative value of -0.2 V. At room temperature

$$\exp(eV_D/kT) = \exp(40 \times -0.2) \approx 3 \times 10^{-4}$$

so

$$I_D \approx -I_s$$

and this will be so for larger negative values of V_D. For this reason I_s is called the *reverse saturation current*.

I now have to consider what happens when a neutral region width is greater than the diffusion length of minority carriers. Because of recombination, the minority carrier distribution will settle to a steady-state profile as shown, for holes, in Fig. 5.7. The initial slope of the profile, at the transition region edge, will determine the current density at that point – and hence everywhere else in the neutral region. Theoretical analysis shows that the tangent to the minority carrier density curve cuts the line representing p_{n0} at a distance L_p from the transition region edge, as shown in Fig. 5.7, so an analysis following the lines of the previous one will result in similar equations, except that instead of

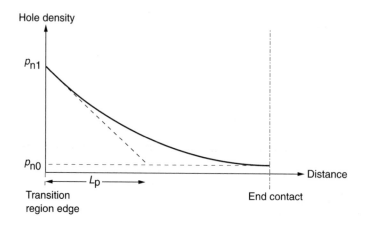

Figure 5.7

Equation (5.8), I_s will be given by

$$I_s = Ae\left[\frac{D_p p_{n0}}{L_p} + \frac{D_n n_{p0}}{L_n}\right] \tag{5.10}$$

Equation (5.9) will again apply, but with this value of I_s.

It is convenient to remember Equations (5.8) and (5.10) together as

$$I_s = Ae\left[\frac{D_n n_{n0}}{(l_n \text{ or } L_p)} + \frac{D_n n_{p0}}{(l_p \text{ or } L_n)}\right] \tag{5.11}$$

but do not be confused by the change of subscript; this occurs because in the one case it refers to the material of the neutral region and in the other it refers to the polarity of the minority carriers, which is the opposite.

The minority carrier profile when the junction is reverse biased gives another indication why the reverse current saturates. I showed earlier that with increasing reverse bias the minority carrier density at the transition region edge rapidly falls to zero and can then fall no further; hence the density gradient – now sloping down towards the transition region – does not change with increasing reverse voltage. Figure 5.8 illustrates the minority carrier profile for

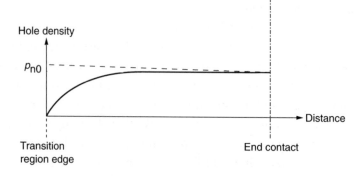

Figure 5.8

any reverse voltage greater than half a volt or so in the case of holes in a wide n-type neutral region.

I can now summarize the flow of current in the forward biased junction. The magnitude of the current is determined by the diffusion of minority carriers away from the transition region edges. As these carriers diffuse away they are replaced by majority carriers from the other side of the transition region and so there must be a drift current of majority carriers to maintain the majority carrier densities. If the neutral regions are wide enough for significant recombination to occur this also will induce a further component of majority carrier drift current to replace the majority carriers as they are eliminated. Voltages associated with the electric fields which arise to drive the majority carrier drift currents are small enough to be neglected.

A final point: since the majority carriers have rearranged themselves to neutralize the electric fields caused by the excess of minority carriers, and thus have a density gradient, why do they not also contribute to the current by diffusing? The answer is that there is no source which would replenish them as they diffused, and thus the equilibrium arrangement will include a small residual electric field to counteract their diffusion. This field will add a drift component to the motion of the minority carriers, but I am assuming that the effect is small enough to be neglected.

Equations (5.9) and (5.11) together give a good indication of the currents which flow for a range of voltages applied to a normal pn junction. Because of the various approximations I have made, and other effects that I have not yet mentioned, there will be circumstances in which the equations do not hold. I shall deal with these as they arise.

The ratio of hole current to electron current

In some applications of pn junctions it is important to know the ratio of hole current to electron current at the junction. In Equation (5.8), $\dfrac{D_p p_{n0}}{l_n}$ represents the proportion of I_s, and hence of I_D, which is contributed by holes, as compared to $\dfrac{D_n n_{p0}}{l_p}$, which represents the proportion contributed by electrons. Hence

$$\frac{I_p}{I_n} = \frac{D_p p_{n0}}{l_n} \bigg/ \frac{D_n n_{p0}}{l_p}$$

$$= \frac{D_p}{D_n} \times \frac{p_{n0}}{n_{p0}} \times \frac{l_p}{l_n}$$

But $p_{n0} \approx n_i^2/N_D$ and $n_{p0} \approx n_i^2/N_A$, so

$$\frac{I_p}{I_n} = \frac{D_p}{D_n} \times \frac{N_A}{N_D} \times \frac{l_p}{l_n}$$

If I had started from Equation (5.10) I should have arrived by a similar process at

$$\frac{I_p}{I_n} = \frac{D_p}{D_n} \times \frac{N_A}{N_D} \times \frac{L_n}{L_p}$$

so I can generalize the result to

$$\frac{I_p}{I_n} = \frac{D_p}{D_n} \times \frac{N_A}{N_D} \times \frac{(l_p \text{ or } L_n)}{(l_n \text{ or } L_p)} \qquad (5.12)$$

5.5   WIDTH OF THE TRANSITION REGION

The following analysis is based on the simplifying assumption that the transition region is entirely depleted of carriers. I shall consider the general case in which a voltage V_D is applied, so that the total voltage across the transition region is $(V_D + \psi)$.

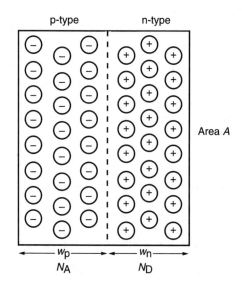

Figure 5.9

Figure 5.9 shows the transition region, with the width on the p side shown as w_p and the width on the n side as w_n. The total quantity of exposed charge on each side of the metallurgical junction must be the same, so

$$N_A w_p A e = N_D w_n A_e$$

and thus

$$\frac{w_p}{w_n} = \frac{N_D}{N_A} \qquad (5.13)$$

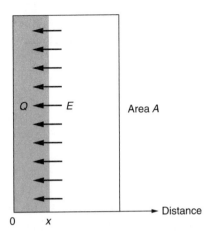

Figure 5.10

The widths are inversely proportional to the doping densities.

Dealing with the p side first, in Fig. 5.10, the electric field at distance x from the transition region edge can be found using Gauss's theorem. The total charge in the shaded area is given by

$$Q = -N_A A x e$$

The minus sign is included because the acceptor ions are negative. All the electric flux is in the transition region and will be perpendicular to the junction, so I can get the flux density from

$$D = Q/A = -N_A x e$$

(the minus sign shows that D is in the opposite sense to positive x).

The electric field $E = D/(\epsilon_0 \epsilon_r)$, i.e. the electric field at x is given by

$$E_x = -\frac{N_A e}{\epsilon_0 \epsilon_r} x \qquad (5.14)$$

Since $E = -\mathrm{d}V/\mathrm{d}x$,

$$\frac{\mathrm{d}V}{\mathrm{d}x} = \frac{N_A e}{\epsilon_0 \epsilon_r} x$$

If I integrate this expression between the limits 0 and w_p I shall get the voltage at the metallurgical junction relative to the transition region p-side edge:

$$V = \frac{N_A e}{\epsilon_0 \epsilon_r} \int_0^{w_p} x\,\mathrm{d}x$$

$$= \frac{N_A e}{\epsilon_0 \epsilon_r} \frac{w_p^2}{2}$$

But what I want is the voltage of the p side relative to the metallurgical junction, which I shall call V_p, and which is the other way round, so

$$V_p = -\frac{N_A e}{\epsilon_0 \epsilon_r} \frac{w_p^2}{2}$$

On the n side, a similar argument yields

$$V_n = -\frac{N_D e}{\epsilon_0 \epsilon_r} \frac{w_n^2}{2}$$

Care is needed to make sure that the signs are correct to give the voltage of the p side relative to the n side. I have tried here to be mathematically rigorous, but generally it is easier to work out the magnitudes of values and rely on physical understanding to ascribe a sign.

Since both sides of the following equation represent the total voltage across the transition region,

$$V_p + V_n = V_D + \psi$$

so, substituting for V_p and V_n, and changing signs,

$$-V_D - \psi = \frac{e}{2\epsilon_0 \epsilon_r} (N_A w_p^2 + N_D w_n^2)$$

Now ψ is always negative, so $-\psi$ is positive and therefore I can usefully replace $-\psi$ by the positive value of the modulus of ψ, i.e. by $|\psi|$, giving

$$|\psi| - V_D = \frac{e}{2\epsilon_0 \epsilon_r} (N_A w_p^2 + N_D w_n^2)$$

Since, from equation (5.13),

$$w_n^2 = w_p^2 \frac{N_A^2}{N_D^2}$$

$$|\psi| - V_D = \frac{e}{2\epsilon_0 \epsilon_r} \left(N_A w_p^2 + \frac{N_A^2}{N_D} w_p^2 \right)$$

$$= \frac{e N_A^2}{2\epsilon_0 \epsilon_r} \left(\frac{1}{N_A} + \frac{1}{N_D} \right) w_p^2$$

so

$$w_p = \frac{1}{N_A} \sqrt{\left[\frac{2\epsilon_0 \epsilon_r}{e} (|\psi| - V_D) \left(\frac{1}{1/N_A + 1/N_D} \right) \right]}$$

Similarly, by substituting for w_p,

$$w_n = \frac{1}{N_D} \sqrt{\left[\frac{2\epsilon_0 \epsilon_r}{e} (|\psi| - V_D) \left(\frac{1}{1/N_A + 1/N_D} \right) \right]}$$

The total width of the transition region is given by

$$w_t = w_p + w_n$$

Adding the two functions and rearranging leads to

$$w_t = \sqrt{\left[\frac{2\epsilon_0\epsilon_r}{e}\left(\frac{1}{N_A} + \frac{1}{N_D}\right)(|\psi| - V_D)\right]}$$

Self assessment test 5.5

For the example junction, find the width of the transition region at room temperature (a) for no applied voltage, (b) for a forward applied voltage of 0.5 V, and (c) for an applied reverse voltage of 0.5 V.

I have already calculated in a previous example the contact potential for this junction at room temperature: it is -0.83 V. Also, $N_A = 5 \times 10^{23}$ m^{-3} and $N_D = 5 \times 10^{22}$ m^{-3}. Substituting values for the general problem,

$$w_t = \sqrt{\left[\frac{2 \times 8.85 \times 10^{-12} \times 12}{1.6 \times 10^{-19}} \times \left(\frac{1}{5 \times 10^{23}} + \frac{1}{5 \times 10^{22}}\right) \times (0.83 - V_D)\right]}$$

$$= 1.7 \times 10^{-7}\sqrt{(0.83 - V_D)}$$

(a) $V_D = 0$ V, $w_t \approx 0.15\ \mu$m

(b) $V_D = 0.5$ V, $w_t \approx 0.1\ \mu$m

(c) $V_D = 0.5$ V, $w_t \approx 0.2\ \mu$m

You will notice from the solutions the order of magnitude of the transition region width – tenths of a micrometre, i.e. 10^{-7} m. You will also notice that forward bias narrows the transition region, while reverse bias widens it.

Self assessment test 5.6

For the example junction, which side of the junction will contain the greater part of the transition region?

Explanation

From equation (5.13),

$$\frac{w_p}{w_n} = \frac{N_D}{N_A} = \frac{5 \times 10^{22}}{5 \times 10^{23}} = \frac{1}{10}$$

so, whatever the bias, 10/11 of the transition region is in the lightly doped n side and 1/11 is in the heavily doped p side.

5.6 ELECTRIC FIELD IN THE TRANSITION REGION

You will see from Equation (5.14) that, based on the assumption that there are no carriers in the transition region, the magnitude of the electric field strength increases linearly with distance from the transition region edge to the metallurgical junction. If you think of electric flux passing from charges in one side of the transition region to opposite charges in the other, the greatest density of flux, and hence the largest value of electric field, must be at the metallurgical junction. Measurements for actual junctions give results similar to Fig. 5.11, which show that the depletion layer edges are not quite so well defined as the simple model suggests.

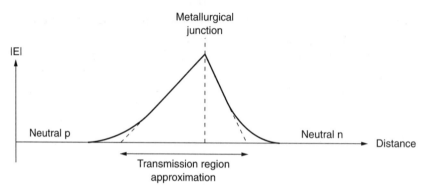

Figure 5.11   Magnitude of the electric field strength with distance through a pn junction – the n side is twice as heavily doped as the p side.

Since, approximately, the electric field strength increases linearly from zero to a maximum and then decreases linearly to zero with distance across the junction, the average value of electric field is half the maximum value. Hence, this average value times the transition region width must give the voltage difference across the transition region. I can write

$$\frac{E_{max}}{2} \times w_t = |V_D + \psi|$$

So

$$E_{max} = \frac{2|V_D + \psi|}{w_t} \tag{5.16}$$

In applying this formula you must remember that the value of ψ is negative.

Self assessment test 5.7

For the example junction, calculate the value of the electric field strength at the metallurgical junction with no bias applied. How will bias affect this result?

Explanation

In previous examples we found that the contact potential for this junction at room temperature is -0.83 V and the width of the transition region for zero bias is $\approx 1.5 \times 10^{-7}$ m. Hence

$$E_{max} \approx 2 \times 0.83/1.5 \times 10^{-7} \approx 10^{7} \text{ V m}^{-1}$$

For forward bias the magnitude of the total voltage across the junction is reduced, so E_{max} is less; for reverse bias the total voltage is increased so E_{max} is greater.

The fact that the electric field strength in a junction can be very high is significant in the design of devices for use at high frequencies. If you wish to calculate how fast the carriers of a reverse current drift through the junction, you cannot use the quoted mobility values; these only apply up to moderate field strengths in the material. At field strengths of around 10^{6} V m^{-1} in semiconductor materials the carrier velocities 'saturate' and any further increase in field strength does not result in a further increase in carrier velocity.

Inspection of Equations (5.15) and (5.16) suggests something that you may already have suspected – that the forward voltage which you can apply to a pn junction must be less than the magnitude of the contact potential. Equation (5.15) indicates that a forward voltage of $|\psi|$ would lead to the elimination of the transition region, and, according to Equation (5.16), the electric field would be zero. If the junction were ideally abrupt there would be an infinite minority carrier density gradient and thus infinite current. Of course, this ignores the simplifying assumptions used to deduce the equations, one of which was that the actual junction current is vanishingly small compared to the balancing drift and diffusion currents in the transition region. Nothing in what has gone before suggests any limit to the reverse voltage that can be applied – there is a limit, which I shall discuss in the next chapter.

5.7 TRANSITION REGION CAPACITANCE

When the voltage across a pn junction is changed, the width of the transition region changes, involving a change of charge content in the transition region on either side of the metallurgical junction – a capacitive effect. The stored charge is not proportional to the applied voltage, so I cannot define a d.c. capacitance as Q/V. However what does prove to be significant is the a.c. or 'small signal' capacitance, defined as dQ/dV.

As in any capacitor, the quantity of charge considered is the increase of positive charge on one side or the increase in negative charge on the other – both being the same – and so this transition region capacitance, C_t, is defined as the rate of increase of transition region charge on either side of the metallurgical junction with applied voltage. Proceeding along similar lines to the analysis in Section 5.5, it is possible to show that, for an abrupt junction,

$$C_t = A \sqrt{\left[\frac{e\epsilon_0\epsilon_r}{2\left(\dfrac{1}{N_A} + \dfrac{1}{N_D}\right)(|\psi| - V_D)} \right]} \qquad (5.17)$$

Again it is clear that this formula does not allow for a forward applied voltage greater than $|\psi|$. The capacitance varies with voltage, being greater for forward than for reverse bias, and falling with increasing reverse bias.

5.8   DIFFUSION CAPACITANCE

Besides the transition region capacitance, there is another capacitive effect in a semi-conductor crystal which contains a pn junction. This capacitance is only significant when the junction is forward biased, and arises because a change in the applied voltage results in a change in the total charge content of the neutral regions.

Formulae to calculate the diffusion capacitance differ depending on the particular arrangement. By way of illustration I shall deduce those which apply if the neutral regions are short, and thus there is negligible recombination other than at the terminals.

Let us look first at the n side of such a junction to which a forward bias has been applied. Figure 5.12 shows the carrier profiles in the neutral region:

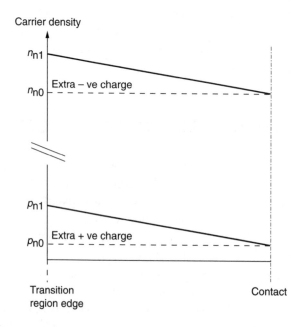

Figure 5.12

the vertical axis is broken because the minority and majority carrier densities have very different magnitudes. Compared to the situation with no applied voltage there is a wedge of extra positive charge (producing the minority carrier density gradient) and a wedge of extra negative charge (the result of dielectric relaxation). When the voltage was applied, the extra holes came via the transition region, and so the positive charge ultimately entered the crystal at the p-side contact, while the extra negative charge entered the crystal via the n-side contact, so although there is nothing here which even remotely resembles a pair of plates, nevertheless there is a capacitive effect. The capacitance can be defined as dQ/dV_D, where Q can be either the extra hole charge or the extra electron charge (they are the same magnitude).

Considering the hole charge, we know that

$$I_p = AeD_p \frac{(p_{n1} - p_{n0})}{l_n}$$

and you should easily see that the extra hole charge

$$Q_p = \frac{(p_{n1} - p_{n0})}{2} l_n Ae$$

Combining these results gives

$$Q_p = \frac{l_n^2}{2D_p} I_p$$

The diffusion capacitance contribution of the n side, C_{dn}, is given by

$$C_{dn} = \frac{dQ_p}{dV_d} = \frac{dQ_p}{dI_p} \times \frac{dI_p}{dV_d}$$

From the previous expression,

$$\frac{dQ_p}{dI_p} = \frac{l_n^2}{2D_p}$$

To get dI_p/dV_D, I need to use the equation, derived in Section 5.4,

$$I_p = \frac{AeD_p p_{n0}}{l_n} \left[\exp\left(\frac{eV_D}{kT}\right) - 1 \right]$$

Differentiating this, we get

$$\frac{dI_p}{dV_d} = \frac{e}{kT} \frac{AeD_p p_{n0}}{l_n} \exp\left(\frac{eV_D}{kT}\right)$$

So, approximately,

$$\frac{dI_p}{dV_d} = \frac{eI_p}{kT}$$

Altogether then,

$$C_{dn} = \frac{l_n^2 I_p e}{2D_p kT} \tag{5.18}$$

By similar reasoning for the p side,

$$C_{dp} = \frac{l_p^2 I_n e}{2D_n kT} \tag{5.19}$$

If the doping is such that the junction current is carried mainly by holes then Equation (5.18) gives approximately the diffusion capacitance; if electrons are the predominant carriers than Equation (5.19) should be used. For junctions with similar doping densities on both sides, results from Equations (5.18) and (5.19) must be added to give the total diffusion capacitance.

For long neutral regions, where all the recombination takes place in the bulk of the material, I cannot, in this case, simply replace neutral region widths by diffusion lengths. It turns out that the following relationships hold:

$$C_{dn} = (t_p I_p)\frac{e}{kT} \quad \text{and} \quad C_{dp} = (t_n I_n)\frac{e}{kT}$$

where t_p and t_n are minority carrier lifetimes.

If a junction has one long neutral region and one short, with similar doping densities on either side, then it may be necessary in order to calculate the total diffusion capacitance to use one of the above formulae together with one of Equations (5.18) and (5.19).

An important thing to note is that diffusion capacitance formulae do not show any dependence on junction area. The diffusion capacitance is, however, proportional to the forward current which flows (this involves the area in the sense that basic properties, independent on the particular specimen, would normally be given in terms of current density).

5.9   NON-ABRUPT JUNCTIONS

The processes by which semiconductor devices are manufactured often results in pn junctions in which the transition from p to n is not abrupt; rather the effective acceptor doping falls over a distance to zero at what is regarded as the metallurgical junction, and from there the donor doping increases with distance into the n side. The general form of the current/voltage equation (Equation (5.9)) is not affected. However, the transition region width is greater than for an abrupt junction between the same levels of neutral region doping and the transition region capacitance is no longer given by Equation (5.17).

It should be easy to see why the transition region will be wider: its width is that required to provide the appropriate amount of uncovered ion charge on either side of the junction. If there are few ions near the metallurgical junction the transition region must extend further back.

Because the average separation of the charges is greater, the transition region capacitance is smaller than for the analogous abrupt junction, and analysis shows that if the doping varies linearly through the transition region, then the value of C_t is inversely proportional to the cube root of $(|\psi| - V_D)$ rather than to the square root, as in the abrupt junction.

The graph of electric field versus distance through the transition region of a graded junction is not triangular at the centre, as in Fig. 5.11, but rounded off.

If the doping variation extends beyond the transition region there will be electric fields in the 'neutral regions' for reasons explained at the end of Chapter 4. This may significantly modify the minority carrier currents.

5.10 SUMMARY

Spanning a pn junction in a semiconductor crystal is a transition region, typically of the order of 10^{-7} m thick, consisting of 'uncovered' acceptor and donor ions. The transition region extends further into a lightly doped material than into a heavily doped material. A forward bias voltage narrows the transition region, while a reverse applied voltage widens it.

The transition region contains an electric field, of maximum magnitude at the metallurgical junction. Beyond the transition region, on either side, are neutral regions no different from the same material in the absence of a pn junction.

With the crystal isolated (not connected into a circuit) there is a contact potential between the two neutral regions, whose magnitude is given by

$$\psi = (kT/e) \ln[n_i^2/(N_A N_D)]$$

and its sign, according to the convention that a voltage is expressed as that of the p side relative to the n side, is negative.

The electrical condition of an isolated crystal of a semiconductor can be illustrated by an energy band diagram, such as Fig. 5.4.

When a constant voltage V_D is applied to a crystal containing a pn junction the current that flows is given by

$$I_D = I_s \left[\exp\left(\frac{eV_D}{kT}\right) - 1 \right]$$

where, provided the doping in the neutral regions is uniform,

$$I_s = Ae \left[\frac{D_p p_{n0}}{(l_n \text{ or } L_p)} + \frac{D_n n_{p0}}{(l_p \text{ or } L_n)} \right]$$

and the ratio of hole current to electron current is given by

$$\frac{I_p}{I_n} = \frac{D_p}{D_n} \times \frac{N_a}{N_D} \times \frac{(l_p \text{ or } L_n)}{(l_n \text{ or } L_p)}$$

The Ls in the above formulae are diffusion lengths, while the ls are neutral region widths; the one to use in each case is that with the smaller value.

For an abrupt junction with uniform doping on either side, the transition region width at applied voltage V_D is given by

$$w_t = \sqrt{\left[\frac{2\epsilon_0\epsilon_r}{e}\left(\frac{1}{N_A} + \frac{1}{N_D}\right)(|\psi| - V_D)\right]}$$

and the proportion of the transition region on each side of the metallurgical junction is inversely proportional to the doping density.

There are two capacitive effects in the crystal: the first, associated with changes in the transition region width with changing applied voltage, is the transition region capacitance; and the second, associated with changes in excess charge in the neutral regions with changing applied voltage, is the diffusion capacitance. For forward bias conditions the diffusion capacitance usually dominates; for reverse bias, although the values of transition region capacitance are smaller than they are for forward bias because there is negligible diffusion capacitance, transition region capacitance dominates. Neither of these capacitances is fixed; the diffusion capacitance is proportional to the forward current, while the transition region capacitance varies inversely with the total junction voltage (contact potential plus reverse voltage) – as the square root of this voltage for an abrupt junction, or the cube root for a linearly graded junction.

5.11 PROBLEMS

A square slice of single-crystal germanium of area 1 mm^2 and thickness 200 μm contains an abrupt junction, parallel to the faces, between uniform doping with aluminium of density 10^{22} atoms m^{-3} and uniform doping with arsenic of density 10^{23} atoms m^{-3}. The aluminium doped region is 5 μm thick, the rest of the slice being the arsenic doped region. Metal contacts, which form surfaces of infinite recombination, are made to the two faces. The relative permittivity of germanium is 16. Take the minority carrier lifetimes in both neutral regions to be 10^{-7} s. Assume room temperature.

1 Calculate a value for the contact potential of the junction.

2 Calculate the diffusion lengths of minority carriers in the two neutral-region materials.

3 Calculate a value for the width of the transition region with an applied forward voltage of 0.2 V and hence deduce the approximate width of the neutral region in the p doped material for this bias.

4 Calculate a value for I_s for the 0.2 V forward bias and hence deduce the junction current for this bias.

5 Find the ratio of the electron current to the hole current through the junction.

6 Estimate the diffusion capacitance for 0.2 V forward bias.

7 Find the transition region width for a reverse bias of 2 V and hence find a value for I_s at this bias. Why is it not the same value as found in question 4?

8 Estimate the transition region capacitance for a reverse bias of 2 V.

6 Semiconductor diodes

In this chapter we move on from underlying physics to the consideration of devices. I shall make some reference, as necessary, to how things are made, but for any detail you will have to look elsewhere (see Chapter 1). The term 'diode', meaning literally 'two electrodes' is a hang-over from the days of vacuum electronic devices; nowadays it is used to refer to any device with two terminals and non-reciprocal characteristics, that is, with a current response which is different depending on which way round the voltage is applied.

It will be necessary frequently to refer to the theoretical characteristic equation for a pn junction

$$I_D = I_s[\exp(eV_D/kT) - 1]$$

which, for brevity, I shall simply call 'the diode equation'.

I shall discuss rectifying diodes, small-signal diodes, voltage reference diodes and Schottky diodes. With the possible exception of the first, all these devices appear most frequently as components of integrated circuits, and you may assume in each case that the semiconductor material is silicon.

In addition to the devices discussed in this chapter there are a number of diodes with different structures and using different semiconducting materials used at very high radio frequencies, and also special diodes used for the emission and detection of infrared and visible light signals.

6.1 RECTIFYING DIODES

A rectifying diode is one whose purpose is to conduct in one direction and not in the other, so as to convert alternating current (generally sinusoidal) into pulsating direct current. The term is usually applied to devices for use with high powers and low frequency – rectifying a voltage derived from the mains in a power supply is typical. The ideal characteristic (that is, current/voltage curve) for such a device would be as shown in Fig. 6.1; there would be no current whatever with applied reverse voltage and zero voltage drop for a forward current, thus no power would be dissipated by the diode.

Figure 6.1 Characteristic of an ideal rectifier.

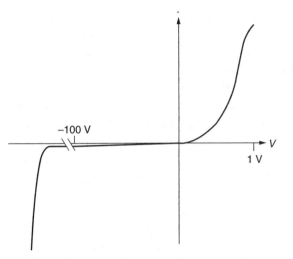

Figure 6.2 Characteristics of a pn junction diode (forward and reverse voltage scales are different).

How does a pn junction characteristic measure up to this? The characteristic of a practical, well-made pn junction which might be used for this purpose is as shown in Fig. 6.2. The forward current/voltage characteristic follows the $\exp(eV_D/kT)$ law up to a value of current at which voltage drop in those parts of the neutral regions in which drift current flows becomes significant. The reverse current at low reverse voltages is very small, as theory suggests, but it will be larger than the I_s value associated with the forward characteristics, and as the reverse voltage is increased the reverse current will increase – at first very slowly, but ultimately catastrophically, as shown. The breakdown voltage, at which the reverse current increases virtually independently of voltage, can be many hundreds of volts.

The mechanisms to account for the reverse characteristic are as follows:

- As the transition region widens, one or both of the neutral regions may become significantly narrower, thus increasing the reverse current.
- There is leakage across the junction at the surface. The junction has to end at a surface, and the abundance of carriers caused by the distortion of the energy band structure, together with the inevitable surface contamination,

leads to an 'ohmic' current (that is, a current proportional to voltage) which is a significant addition to the junction reverse current. This leakage current is minimized by 'passivating' and 'encapsulating' the surface. Passivating involves coating the silicon with a material to which the 'dangling valence bonds' at the surface can attach: silicon oxide (silica) is appropriate. Encapsulating involves sealing to keep out moisture, etc. Both effects are sometimes achieved by coating the surface round the junction with a glass (glass includes silica as a constituent).

- The breakdown is caused by avalanching in the transition region. There is, as we have seen in Chapter 5, a strong electric field in the junction, and this field accelerates carriers in a reverse current. Imagine carriers being pushed by the field through the transition region and making collisions with the silicon atoms from time to time on a random basis, but with a constant average distance between collisions called the 'mean free path'. Take an electron as an example. As the voltage is increased, and thus the electric field increases, a value will be reached at which the electron, between interactions with silicon atoms, acquires enough energy from the field to ionize a silicon atom, that is, to knock a valence electron out into the interatomic space as a conduction electron. A hole is also released. There are now two electrons, plus a hole, which are accelerated (the hole in the opposite direction to the electrons) to the next ionizing collisions and so on. Energetic holes will similarly ionize silicon atoms. Obviously, not all collisions will be 'head-on' and so not all will be ionizing, but as the field at any point increases, the energy acquired between interactions increases, so the chance of a collision causing ionization will increase and so there will be carrier multiplication, increasing with voltage.

Equations (5.15) and (5.16) of Chapter 5 together indicate that, for given doping, the maximum electric field value increases approximately as the square root of the voltage. Significant avalanche multiplication starts to occur in silicon at field strengths of about 5×10^7 V m^{-1}. Once the peak field strength exceeds the avalanche multiplication value, further increase in voltage will increase the length of that part of the transition region in which avalanching can occur; this again causes the current to increase. In an abrupt junction, as we have seen in Chapter 5, the field strength is a maximum at the metallurgical junction and its magnitude, for a given voltage, is smaller the greater the junction width; hence, for a high reverse breakdown voltage a wide transition region is required.

As the reverse current increases due to avalanche multiplication the power dissipated in the device – indicated by current times the voltage – becomes significant and the device gets hot, leading to greater current, leading to more heat and so on, a process known as 'thermal runaway' which rapidly results in destruction of the device. Until thermal runaway takes hold, the breakdown is reversible; the current will fall back if the voltage is reduced.

Let me now outline some of the problems associated with designing a rectifying diode. I want good contacts to the p and n sides: this is usually achieved
by very heavy doping of the p and n material to which a metallic contact is
made. I want reasonably heavy doping of the p and n materials up to the junction so that the neutral regions are good conductors, and the voltage drops in
them at the highest rated currents for the device are insignificant. This is my
first problem: heavy doping means a narrow transition region and thus a low
reverse breakdown voltage.

Suppose I were to reduce the doping density, but increase the area of the
device to reduce the resistance of the neutral regions? This gives rise to
another problem which is associated with the manufacture of semiconductor
devices – the problem that material faults are randomly distributed over the
area of a slice of semiconductor from which a chip is made. In the manufacturing process slices are divided up into chips and those with faults are rejected.
The larger the areas of the chips, the greater will be the percentage with faults,
and thus the lower the yield. On the other hand, if a chip area is too small it
will be difficult to conduct away the heat generated by the forward current:
everything is a compromise.

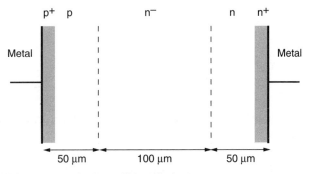

Figure 6.3   Suitable structure for a rectifying diode

A design which is often adopted to give high reverse breakdown voltage
with low neutral region forward resistance is that shown in Fig. 6.3. Typical
widths are shown. An appropriate area would be, say, 5 mm^2. The p$^+$ and n$^+$
regions have doping densities of the order of 10^{26} m^{-3}, the p and n regions
about 10^{24} m^{-3}, and the n$^-$ region about 10^{20} m^{-3}. The junction is between the p
and n$^-$ materials, and virtually all the transition region is in the n$^-$ material. It
is intended that the transition region should just extend right across the n$^-$
region at the design maximum reverse voltage.

Self assessment test 6.1 _____

Calculate, for the dimensions and values given, the reverse voltage at which
the transition region just fills the n$^-$ material.

Answer

From Equation (5.15) of Chapter 5,

$$-V_D \approx w_t^2 \times \frac{e}{2\epsilon_0\epsilon_r(1/N_D)}$$

$$= \frac{(10^{-4})^2 \times 1.6 \times 10^{-19}}{2 \times 8.85 \times 10^{-12} \times 12 \times 10^{-20}}$$

$$\approx 750 \text{ V}$$

Self assessment test 6.2

Now calculate the maximum field strength at the reverse voltage obtained in test 6.1.

Answer

From Equation (5.16) of Chapter 5,

$$E_m \approx 2V_D/w_t$$

$$= 2 \times 750/10^{-4}$$

$$= 1.5 \times 10^7 \text{ V m}^{-1}$$

The solution to test 6.2 indicates that avalanche current multiplication should not be a problem.

What about the forward voltage drop? The resistivity of silicon doped with 10^{20} m^{-3} donors is $0.4 \ \Omega$ m, which would be expected to produce a significant voltage drop over a distance of, say, $80 \ \mu$m for a drift current of several amps. If the n^- material is pure enough for the minority carrier lifetime to be very high, say 10^{-5} s, then the injected holes will diffuse to the n material, and the resistivity of the n^- material will not be important, but this can have an attendant disadvantage. When the applied voltage is reversed, stored charge in the diode has to be eliminated before it settles to its steady-state reverse-biased condition. If the minority carrier lifetime is long, excess holes in the n^- material will flow back into the p material rather than recombining, thus producing a transient pulse of reverse current. Depending on the frequency and the application, a shorted minority carrier lifetime in the n^- material might be necessary, but in any case the large injection of holes associated with a large forward current will draw a similar number of electrons into the lightly doped material, thus increasing its effective conductivity.

6.2 SMALL-SIGNAL DIODES

Diodes are used in high frequency analogue circuits for such purposes as signal detection, frequency discrimination and mixing. Normal silicon pn junctions, of small area, are appropriate for these applications except for the very highest frequencies, in the GHz range.

The currents involved are generally a few milliamps or less, and heat dissipation and reverse voltage breakdown do not generally present design problems. Two parameters which are significant are slope resistance and a.c. capacitance.

A diode may have a direct bias current with a small a.c. voltage superimposed. The a.c. current produced in the diode by this a.c. voltage is determined by the slope resistance and the a.c. capacitance, which are effectively in parallel.

Assuming that the forward-biased diode obeys the diode equation, for a significant constant forward voltage I can approximate this as

$$I_D = I_s \exp(eV_D/kT)$$

Differentiating with respect to voltage yields

$$\frac{dI_D}{dV_D} = I_s\frac{e}{kT}\exp(eV_D/kT) = I_D\frac{e}{kT}$$

Using the standard symbol for the slope resistance, r_e,

$$dI_D/dV_D = 1/r_e = I_D(e/kT)$$

$$r_e = \frac{(kT/e)}{I_D}$$

At room temperature, $kT/e = 25$ mV, so that, at room temperature,

$$r_e = \frac{25}{I_D(\text{in mA})}\ \Omega$$

This formula normally gives an accurate representation of the behaviour of a diode with a bias current in the milliamps range; however, for lower currents the slope resistance is greater, tending towards twice the value that the above formula would indicate. The reason for the deviation is that some recombination of carriers in the transition region takes place and this was ignored in developing the diode equation. The effect cannot, however, be ignored at very low currents, and it causes the exponential term in the equation to move towards $(eV_D/2kT)$.

It will be useful to establish the order of diode voltage associated with a current of 1 mA in a small-signal silicon diode. I shall assume an area of 1 mm^2, one part per million of dopant on either side of the junction – that is,

5×10^{22} m^{-3} – and long neutral regions with minority carrier lifetimes of 10^{-7} seconds, giving $L_n = 20\ \mu$m and $L_p = 11\ \mu$m, as calculated in Chapter 5.

Self assessment test 6.3 _____

Calculate I_s for the diode described in the previous paragraph.

Answer

Using Equation (5.10) of Chapter 5, the result is $I_s \approx 10^{-13}$ A.

Self assessment test 6.4 _____

Hence find the forward voltage required for a current of 1 mA at room temperature.

Answer

From Equation (5.9) of Chapter 5,

$$eV_D/kT = \ln[(I_D + I_s)/I_s]$$
$$V_D/0.025 = \ln(10^{-3}/10^{-13}) = \ln(10^{10})$$
$$V_D = 0.58\ \text{V}$$

The results of self-assessment tests 6.3 and 6.4 are typical.

Usually, one will want to keep the a.c. capacitance as low as possible – particularly the diffusion capacitance, which is the more significant for a forward-biased diode. The value of this is directly proportional to the bias current.

The use of a small-signal diode as a mixer relies on the non-linearity of its characteristic.

6.3   VOLTAGE REFERENCE DIODES

The ideal characteristic of a voltage reference diode, as usually employed in electronic circuits, is that shown in Fig. 6.4. Any current, constant or varying, flowing in such a diode would, so long as it were in the right direction, result in exactly the same voltage between its ends.

The breakdown part of the characteristic of a specially designed pn junction is often used for this purpose. Such diodes are usually called Zener diodes, although the title is not always appropriate as we shall see.

I introduced the idea of avalanche breakdown in Section 6.1. How could one

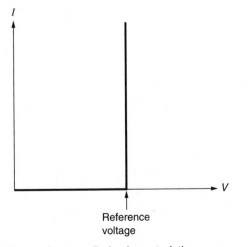
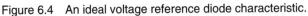

Figure 6.4 An ideal voltage reference diode characteristic.

design a diode so that avalanche breakdown would occur at a suitable low voltage, say 10 V? It would be necessary to use heavy doping so that the maximum electric field reaches 5×10^7 V m^{-1} at a little below 10 V. Assuming equal doping on either side, doping densities of about 2×10^{23} m^{-3} are required – exact values would have to be found by trial and error.

The reverse characteristic of the diode described above would look as shown in Fig. 6.5. A maximum current is specified – perhaps of the order of 20 mA – below which heat liberated in the diode can be dispersed, so that there is no overheating. The voltage is not quite constant as the current changes: above the knee the slope resistance is typically 10 or 12 ohms or so, but somewhat lower at higher currents. The characteristic changes slightly with temperature, moving to higher voltages as the temperature increases, but the change in voltage is only about 0.01 per cent per Kelvin, so the device is useful as a fixed voltage reference.

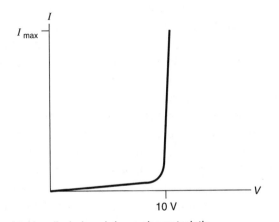

Figure 6.5 An avalanche diode breakdown characteristic.

Suppose the doping densities were increased to over 10^{24} m^{-3}. The avalanche breakdown voltage would now be less than 1 V. However, with this level of doping, the transition region is very narrow so that current multiplication does not take place when the breakdown voltage is exceeded – there is not enough distance for a significant number of ionizations to occur. If the voltage is increased to a value at which the maximum field strength is 1.2×10^8 V m^{-1}, another breakdown mechanism occurs known as Zener breakdown. In this mechanism, ionization is produced by electrons being pulled directly out of valence bonds by the electric field. The characteristic of a Zener breakdown is similar to that of an avalanche breakdown, as shown in Fig. 6.5, with the following differences: the breakdown voltage will be lower; the slope resistance is generally rather smaller; and the breakdown voltage decreases with increasing temperature.

Below 4 V, breakdown is purely by the Zener mechanism; above 6 V, avalanching is entirely responsible; between 4 and 6 V, there is a mixture of the two effects and, since their temperature coefficients have opposite signs, at about 5 V a reference diode can be made with no variation of breakdown voltage with temperature.

[***Note*** You should note that there is a special circuit, not involving diode breakdown, which is often used as a voltage reference in integrated circuits. This is referred to, somewhat loosely, as a 'bandgap diode', although it is not in fact a diode at all, except in so far as it has two active terminals.]

6.4   SCHOTTKY DIODES

A Schottky diode uses a metal–semiconductor junction, usually aluminium–silicon. Up to now I have assumed that all contacts between metals and semiconductors are ohmic, i.e. they conduct equally well either way. To make sure that this is so when required, the semiconductor material at the contact is very heavily doped so that the crystal structure is very distorted and there is effectively no energy gap, but there are plenty of carriers with very short lifetimes. If any sort of transition region is formed it is so narrow that carrier can penetrate it by a process known as 'tunnelling'. Such a contact has been described previously as a surface of infinite recombination.

However, junctions carefully constructed between metals and semiconductors can, if the energy band structures are appropriate, have properties rather like a pn junction; aluminium to n-type silicon is one of these.

Aluminium, although a metal, is trivalent and so can form a reasonable bond to the surface atoms of a crystal of n-type silicon if it is laid down carefully. The initial condition at the moment of formation of the junction is illustrated in Fig. 6.6. The reason that aluminium behaves as a metal is that its bands in

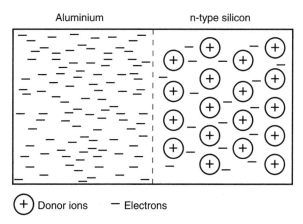

Figure 6.6   Charges in an aluminium/n-type silicon junction at the instant of its formation.

the energy band structure overlap, but the result is similar to that described for copper in that there is a large population of conduction electrons. The material is, of course, overall neutral although only the conduction electrons are indicated in Fig. 6.6.

Looking at Fig. 6.6 you might think that electrons will diffuse from the aluminium to the silicon, but the reverse is in fact the case. To see why, one must consider the band diagrams of the two materials. Figure 6.7 shows the band diagrams for the materials when they are separate. Aluminium has a work function of 4.25 eV, while that of n-type silicon is about 3.7 eV. When the two are combined, with reasonable crystalline continuity, the more energetic conduction electrons in the conduction band of the silicon will flow into the aluminium, while those electrons in the aluminium that are at higher energies than the silicon valence electrons cannot enter the silicon because they have energies in the silicon band gap. The equilibrium condition is with the Fermi

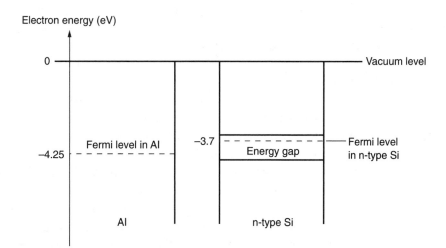

Figure 6.7   Energy band diagrams for aluminium and n-type silicon.

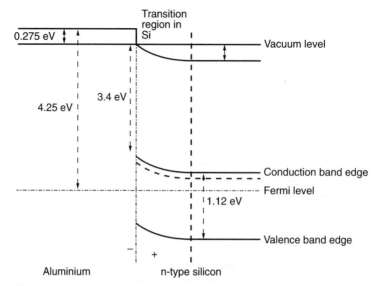

Figure 6.8   Energy band diagram for an aluminium/n-type silicon junction in equilibrium

levels aligned as shown in Fig. 6.8: the silicon has become positively charged and the aluminium negatively charged. Electrons in the aluminium form a negative layer on the surface, as one would expect of a metal, but in the silicon the positive layer is exposed donor ions; a transition region is formed which corresponds to half that in a pn junction.

Forward bias is that bias which opposes the contact potential, i.e. positive to the aluminium and negative to the silicon. There are no holes in the aluminium, so that forward current consists entirely of electrons that diffuse across the transition region from the silicon into the aluminium. No density gradient of electrons can arise in the aluminium, since there is only one type of carrier, and the current in the aluminium will be entirely drift current. The voltage drop in the aluminium is negligible, and the current is controlled by the diffusion of electrons across the transition region. Because the current is diffusion-controlled, the usual diode relationship

$$I_D = I_s[\exp(eVD/kT) - 1]$$

holds. The flow of electrons through the silicon neutral region is also entirely due to drift, and if the doping density is not too low the voltage drop in the silicon neutral region for currents of a few milliamps is also negligible.

Reverse bias produces a saturation current of holes in the silicon neutral region, diffusing to the transition region and accelerating through it to the junction, where they recombine (plus a few holes generated in the transition region, since it is not in equilibrium). Also, a very few electrons in the aluminium have high enough energy to enter the silicon at allowed levels. Looking again at Fig. 6.8, you will see that the electric field is in a direction which would push electrons from the aluminium to the silicon if there were energy levels in the silicon for them to enter. Remembering that the most energetic conduction electrons in the aluminium will be only a little above the Fermi level, it is the lack of available energy levels in the transition region that keeps the electrons out. Thus the transition region forms what is known as a 'Schottky barrier' to electrons; there are appropriate energy levels available in the silicon neutral region for the more energetic aluminium electrons, if they could reach there. If the doping of the silicon is very heavy, the transition region, and hence the Schottky barrier, is so narrow that electrons can penetrate it by 'tunnelling' (a process that can only be explained by quantum mechanics) and thus reach the silicon neutral region. In this case large reverse currents flow and the junction is non-rectifying.

Detailed analysis shows that, for the same junction area, the saturation current of a Schottky diode is about five orders of magnitude higher than in a pn junction, with similar doping densities to that of the n-type silicon, i.e. $\approx 10^{-8}$ A. Using this value of I_s in the diode equation gives a value for the voltage, for a current of 1 mA at room temperature, of about 0.3 V.

Schottky diodes often appear as components in silicon switching circuits. Use is made of the lower forward voltage compared to a silicon pn junction and the fact that the current stops very quickly when the voltage is removed. They have this latter property because no excess charge is stored in the diode in the form of diffusion gradients, so virtually no charge has to be removed at turn-off.

6.5   SUMMARY

Rectifying diodes are used in the process of delivering significant d.c. power from an a.c. source. Silicon pn junction diodes used for this purpose have characteristics that depart from the simple diode equation at high forward currents and at reverse voltages. At high forward currents the voltage drop across the diode, and hence the energy dissipated, is increased by the resistance of the neutral regions. At reverse voltages the reverse current is increased by surface leakage and by the onset of avalanche carrier multiplication in the transition region. This latter effect leads, at high enough voltages, to catastrophic breakdown.

Surface leakage effects are minimized by surface passivation and encapsulation. The breakdown voltage is set at an acceptably high level by designing for a wide transition region.

The structure shown in Fig. 6.3 gives good contact for the diode leads, a high reverse breakdown voltage and low forward current resistance. Virtually all the transition region is in the n⁻ region and just fills it at the design maximum reverse voltage. Injection of carriers into the n⁻ region with forward voltages results in low resistance at high forward currents.

Small-signal silicon pn junction diodes obey the simple diode equation except at very low forward currents. The slope resistance for a given bias current – in the mA range – is given by

$$r_e = \frac{25}{I_D(\text{in mA})} \, \Omega$$

Typically, for a small-signal silicon diode

$$I_S \approx 10^{-13} \, \text{A}$$

and for a current of a few milliamps

$$V_D \approx 0.6 \, \text{V}$$

Voltage reference diodes make use of avalanche breakdown and Zener breakdown. At the low voltages and currents at which they operate, the amount of heat generated is small, so that thermal runaway does not occur and the breakdown is reversible. A typical characteristic is shown in Fig. 6.5. For reference voltages above 6 V, the mechanism is the avalanche effect; below 4 V, it is the Zener effect; and between 4 and 6 V, it is a combination of the two. The temperature coefficient of the reference voltage is positive for avalanche breakdown and negative for Zener breakdown; a diode designed for use around 5 V has a reference voltage which does not vary with temperature.

Schottky diodes use metal–semiconductor rectifying junctions and are named after the man who first investigated the effect. For an aluminium to n-type silicon junction, forward bias occurs with the aluminium positive and the silicon negative, and results in the injection of electrons from the silicon into the aluminium. For forward current there is no diffusion except across the transition region, where the field direction opposes the flow of electrons; current in the rest of the device flows by the mechanism of drift. There is no significant charge storage in the diode, which makes it suitable for switching and for very high frequency applications. The device obeys the diode equation, but values of I_s are in the nanoamps range so that forward voltages for currents of a few milliamps are typically ≈ 0.3 V.

6.6 PROBLEMS

1 A silicon pn junction diode has uniform doping densities of 10^{22} m^{-3} on each side of the junction. The lengths of the p and n regions from the contact to the junction are each $100 \, \mu m$ and the junction area is 5 mm^2. The diffusion length for electrons in the p material is $30 \, \mu m$ and for holes in the n material is $15 \, \mu m$. Estimate, at room temperature, (a) the reverse breakdown voltage, (b) the voltage for a forward current of 1 A, taking account of neutral region resistance.

2 A small-signal silicon pn junction diode has an I_s value of 5×10^{-14} A. The doping density on the n side is 100 times that on the p side. The length of the p material from the contact to the junction is $50\ \mu$m and the minority carrier lifetime in the p material is 2×10^{-7} seconds. Estimate, at room temperature, (a) the range of forward currents for which the diode voltage lies between 0.6 and 0.7 volts, (b) the corresponding range of values of the slope resistance r_e, and (c) the corresponding range of values of the diffusion capacitance C_d.

7 Transistors

In this chapter I concentrate on explaining the basic action of the two most common types of transistors. I have left the mathematical analysis for the next two chapters, and so there are no numerical problems set.

Transistors are basically three-terminal semiconductor devices. The name 'transistor' is derived from 'transfer resistor', but it is not very relevant to the way the devices are used in modern circuitry.

The basic idea behind all transistors is illustrated in Fig. 7.1, in which the terminals are labelled 1, 2 and 3. The device is connected into a circuit at terminals 1 and 2: its properties in that circuit are determined by an input to terminal 3. It is desirable that the input should not take any power from the circuit producing the controlling signal, so ideally, the input could have one of two properties: either the input impedance could be zero and the controlling signal a current, or the input impedance could be infinite and the controlling signal a voltage applied between terminals 3 and 1.

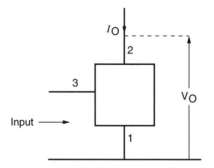

Figure 7.1

There are various possibilities for the circuit property that is controlled by the input. The one to which transistors approximate over a wide range of terminal 2 voltages is as follows: for any applied voltage (but with only one polarity), the current flowing is determined solely by the input signal value.

The earliest transistors were used in such a way that the input approximated to current control; this use could be explained in terms of an input current into a low resistance generating an output current apparently from a high resistance – hence the name.

Now, however, one type of transistor, the field effect transistor, can only be used as a high input impedance device, while the other, the bipolar junction transistor, is usually used in a way that is better considered as voltage rather than current control. (You will see, however, in Chapter 9 that when a bipolar junction transistor is used as a switch, the controlling input parameter is usually a current.)

7.1 CHARACTERISTIC CURVES

The properties of a transistor can usefully be represented by a set of curves known as 'output characteristics'. These show output current against output voltage for a set of different values of the input voltage. An ideal characteristic, for the properties described in the previous section, is that shown in Fig. 7.2. Equal increments in the input terminal voltage cause the horizontal output lines to increase in current value by equal amounts, while the input voltage drives no current into the device.

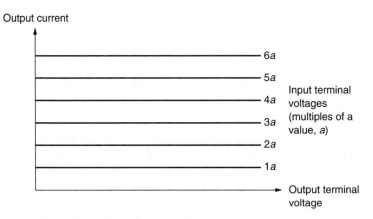

Figure 7.2   Ideal transistor output characteristics.

Such an ideal device has the electrical equivalent circuit of Fig. 7.3, provided that the current I_0 and the voltage V_0 have the relative directions shown (these relative directions ensure that energy is absorbed by the constant current generator rather than given out by it). The parameter g_m is called 'mutual conductance', and simply relates output current to input voltage; its units are amperes per volt (A V^{-1}).

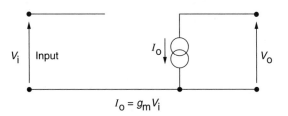

Figure 7.3   Equivalent circuit of an ideal transistor.

You will find the uses of such a device explained in any book on electronics theory, but it should be clear that if the input voltage is varied (though not allowed to reverse in sign), the output current will follow the variations and the device can be used for analogue signals. On the other hand, it can also be used as a switch between a specified input voltage, giving a specified output current, and zero or negative input voltage, giving zero output current.

7.2 MOSFETs

This acronym stands for metal-oxide-silicon field effect transistor, the name indicating the structure. In fact silicon oxide is not always the insulator, nor is silicon always the semiconductor, so the more general term 'insulated gate field effect transistor' (IGFET) is sometimes used, but MOSFET seems to have stuck. It implies a very successful construction technique, known as planar technology, in which devices are made by taking a slab of semiconductor material, called a substrate, and processing it in various ways from one surface only.

[*Note* There is an earlier form of field-effect transistor known as a junction field effect transistor, or JFET, but I shall not discuss these.]

The idea for the field effect transistor is very simple, and was thought up early in the history of semiconductor development; however, the technical problems involved in making it took some time to overcome. There are several different types of MOSFET, all working in a similar way but with detailed differences of semiconductor dopings and voltage polarities, so I will start with one and explain its action; the others can then be described by analogy.

An n-channel enhancement mode MOSFET

Figure 7.4 shows the structure of the device. Into the surface of a substrate of p-type silicon, two n^+ regions have been formed; these are called the 'source'

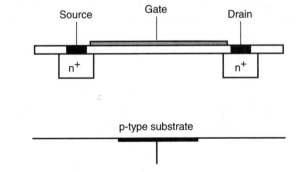

Figure 7.4   Structure of an n-channel MOSFET.

and the 'drain'. The whole surface has been covered with silicon oxide, with two contacts through the oxide to the source and drain; and over the oxide, between the source and drain, is a sheet of conducting material called the 'gate'. It is important that the gate just overlaps the source and drain regions. There is also a contact to the substrate.

The interface between the oxide and the silicon has certain properties that I shall ignore for the moment because they further complicate my explanations. They will be considered in Chapter 8.

The gate is the controlling electrode. If, with no connection to the gate, a small voltage is applied between the source and the drain, positive to the drain let us say, then only a minute current will flow. This current is very small because the pn junction between the drain and substrate is reverse-biased and that between the substrate and the source is only slightly forward-biased.

Now suppose a small positive voltage is applied between the gate and the substrate, so that the gate becomes positively charged: effectively the gate and substrate form a capacitor with the oxide as dielectric. In the substrate, the negative charge next to the oxide forms because the holes move back to expose the acceptor ions, as in a pn junction. There is a difference, however: the minority electrons in the substrate, when they by chance enter the depleted region, are swept by the field towards the oxide; but they cannot cross to the gate. A layer of electrons forms under the oxide with a gradient which just counterbalances the drift of electrons towards the oxide. So, the negative charge under the oxide is part depletion layer and part electron accumulation. Similarly, the holes cannot be entirely excluded from the depleted region because there has to be a diffusion of holes down the gradient created. Equations developed at the end of Chapter 4 will apply, and can be written here as

$$V_{cs} = (kT/e) \ln(p_s/p_c)$$

and

$$V_{cs} = (kT/e) \ln(n_c/n_s)$$

where V_{cs} is the voltage of the surface under the oxide with respect to the substrate, the subscript 'c' referring to the surface under the oxide and 's' to the bulk of the substrate. As the gate voltage is increased, n_c will increase, until a value of V_G is reached at which n_c is the same as the acceptor density. In this state, p_c has the value that the minority carriers have in the bulk of the substrate; in other words there has been an inversion near the oxide, with the electrons having the density of majority carriers and the holes that of minority carriers. The gate voltage at which this occurs is called the 'threshold voltage' and above this voltage an 'n channel' is assumed to exist.

Figure 7.5 shows part of a cross-section. The total gate–substrate voltage is the sum of the voltage across the oxide plus the voltage across the depleted

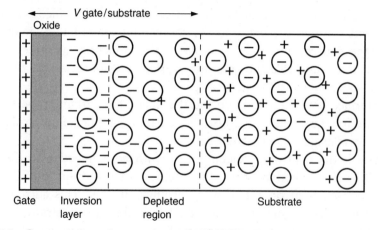

Figure 7.5   Cross-section of an n-channel MOSFET. The relative amount of positive charge on the gate would be greater than shown.

region. Detailed analysis shows that the density of electrons near the surface of the silicon under the oxide is an exponential function of the gate voltage. Now it is a curious property of exponential functions that, although they do not have discontinuities, they can be approximated to as shown, for this case, in Fig. 7.6. The approximation is that below V_T, n_c is negligible, and above V_T, n_c increases linearly with $(V_G - V_T)$. Where you set the threshold value for an exponential function depends on the range of values of the function in which you are interested; for the purpose of simple analysis of the MOSFET, it proves appropriate to take the threshold voltage as given in the previous paragraph.

What happens at the source and drain? Figure 7.7 illustrates the region near the drain junction. Near the surface at the junction the field due to the positive charge on the gate tends to neutralize that due to the negative charge in the p

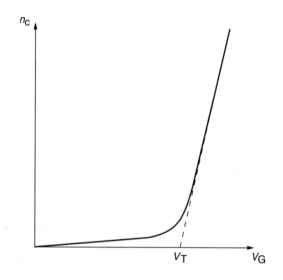

Figure 7.6

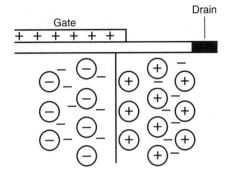

Figure 7.7

side of the depletion layer. Furthermore, since there is no longer a large population of holes on the p side there is no longer a large diffusion gradient of
holes to hold back. Electrons diffuse from the drain into the substrate, and
from the source into the substrate for similar reasons, and although it is the
equilibrium between the diffusion and drift of minority carriers perpendicular
to the gate in the substrate that demands the existence of the inversion layer,
the electrons to form that layer actually come mainly from the source and
drain. In equilibrium, with a positive gate voltage above the threshold,
although there must be some contact potential (and thus some transition
region) between the channel and the drain and between the channel and the
source (unless the channel carrier density has reached the majority carrier densities in the source and drain), there are plenty of carriers in the channel to
carry a drift current. When a small voltage, positive to the drain, is applied
between drain and source, electrons enter the channel from the source, but no
diffusion gradients are set up because there are no holes to neutralize the large
fields that would be created. The electrons drift to the drain junction, where
they are swept into the drain by the junction field.

Since conductivity is proportional to carrier density, you might conclude that
linear variations in the gate voltage, above the threshold value, will simply
cause approximately linear changes in total channel resistance. This would
serve as a controlled output circuit property, but because of the way the voltages are applied, this is not the case. The substrate is normally connected to
the source, so that gate–substrate voltages are gate–source voltages. The reasons for this are firstly that in an integrated circuit the substrate is shared by
many devices and so it is easier to make its potential the common potential,
and secondly, by so doing, the source–substrate junction can never be forward-
biased and current cannot flow directly from the source into the substrate.
Similarly, the voltage applied to the drain is always positive, so that the
drain–substrate junction is not forward-biased.

The effect that I now want to describe is best illustrated with an example.
Suppose the channel threshold voltage is +2 V, the gate–source voltage is set
to +6 V and the drain–source voltage is set to +3 V. There is a voltage drop

along the channel from the drain to the source consistent with the current in the channel. As explained in Chapter 2, this voltage is associated with fields in the channel produced by an appropriate distribution of carriers in the channel and not with electric fields outside the channel. The gate–channel voltage, and the field producing the channel, drops in value from the source end to the drain end. The gate–channel voltage is 6 V at the source end and 3 V (6 V − 3 V) at the drain end.

The distribution of voltage along the channel is not linear. Because the voltage between gate and channel is nearer to the threshold voltage at the drain end than at the source end, the channel resistance is greater at the drain end and so more of the voltage is dropped at the drain end. Figure 7.8 indicates that, moving from the drain, the first volt is dropped over one-fifth of the channel length, while the last volt drop is over almost half the channel length.

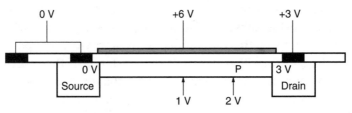

Figure 7.8

Suppose the drain voltage is now increased. When it reaches 4 V, the gate–channel voltage is down to the threshold value at the drain end. Calculations show that almost all the extra volt is dropped down the last fifth of the channel length – the voltage at the point P in Fig. 7.8, a point four-fifths of the channel length from the source, has only risen by about 0.2 V. This has an interesting consequence. The resistance of any part of the channel depends on the voltage at that place. If, as the drain voltage increases, most of the channel's voltage does not change much, then the resistance of the channel over that section does not change much. The voltage at the point P has not changed much, and the resistance from P to the source has not changed much, so the channel current cannot have changed much.

Suppose that the voltage of the drain is increased further, to say 5 V. The gate–channel voltage at the drain end is now well below the threshold and the channel is assumed to be non-conducting. This, however, is incorrect, because the threshold is only an approximation. What actually happens is that the resistivity of the channel rises very rapidly as the gate–channel voltage drops below the threshold value, so that all extra voltage – beyond 4 V for our example – is dropped down a very short high-resistance section of the channel next to the drain. The jargon description is that the channel is 'pinched off' over that section. All the rest of the channel has its voltage distribution almost unchanged, so the channel current changes very little.

If the gate voltage were set at a lower value, pinch-off would occur at a lower drain voltage and the 'saturation current' would have a lower value. The shape of a set of output characteristic curves is shown in Fig. 7.9; the current values are shown as positive because conventional positive currents flow into the device.

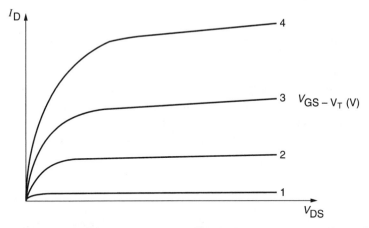

Figure 7.9   n-channel MOSFET characteristics (V_{GS} is the gate–source voltage; V_{DS} is the drain–source voltage).

Other MOSFETs

It is possible to arrange for there to be a layer of fixed positive charge at the oxide to p-substrate interface (in fact a certain number of such positive charges occur naturally, as we shall see in Chapter 8). In this case, an n channel is formed with zero gate voltage, and to pinch off the channel a negative gate voltage is required. The characteristics of a device made in this way are similar to those described in the previous section except that the range of gate control voltages is from negative through zero to positive: it is called an n-channel *depletion* mode MOSFET.

If an n-type substrate is used then a p-channel device results. The drain and source regions will be formed as p^+ material and the required range of drain voltages is negative. The channel currents will be conventionally negative. An enhancement version requires negative gate voltages, while a depletion type, created by laying a fixed negative layer under the oxide, uses both positive and negative gate voltage values.

7.3   BIPOLAR JUNCTION TRANSISTORS (BJTs)

In the days before field effect transistors could be produced these devices were simply called junction transistors to differentiate them from point contact transistors, a structure which is now defunct. To avoid confusion with JFETs the term 'bipolar' was introduced on the grounds that both types of carrier take some part in the flow of current.

A BJT consists of two pn junctions back-to-back with the distance between the two junctions being only a few microns. The structure can be either 'npn' or 'pnp'. I shall start by describing a typical npn transistor.

A silicon planar npn transistor

Figure 7.10 shows the structure of a silicon planar npn transistor. Here, the silicon oxide serves only to seal and passivate the surface. The structure is actually produced by starting with an n-type substrate, which will act as the 'collector', and over-doping with acceptors to form the 'base' region and then again with donors to form the 'emitter'. The 'active' part of the device, where the transistor action occurs, is between the dotted lines.

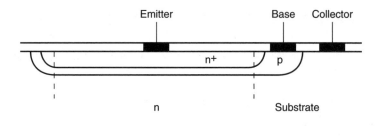

Figure 7.10   A silicon planar npn transistor.

If a voltage is applied between the base and the emitter, so as to make the base sufficiently positive to the emitter, then a current crosses the emitter–base junction. Generally, this current is required to be of the order of milliamps, and the base–emitter voltage to produce this is somewhere in the region of 0.6 to 0.7 V. The emitter is more heavily doped than the base – it could be up to 1000 times – so virtually all the emitter–base current consists of electrons crossing from the emitter to the base.

I am now going to assume that a voltage has been applied between the base and the collector so as to make the collector positive to the base, i.e. to reverse bias the base–collector junction. The diffusion length of minority carriers in the base is substantially greater than the distance between junctions over the active region, so the electrons diffuse away from the emitter until they reach the collector junction transition region, where they are swept into the collector by the transition region field. Thus electrons which are drawn into the base by the voltage applied to the base leave the device not via the base lead, but via the collector lead.

If the transistor behaved exactly as described above, then the current flowing into the collector (remember that electrons flowing out of the collector lead corresponds to conventional current flowing into the collector lead) would be independent of the collector voltage, so long as the base–collector junction remained reverse-biased. The only current flowing in the base lead would be that required to change the charge distribution for the changing diffusion gradient and transition region widths as the base voltage was changed – a capacitive current. However, there are a number of factors which result in a finite resistive current flowing in through the base lead. These will be discussed in Chapter 9.

Let us now deduce the shape of an output characteristic curve for this device with the base–emitter voltage set to, say, +0.6 V. Figure 7.11 shows the physical layout of the active region of the transistor, with the transition region edges in the base labelled E and C.

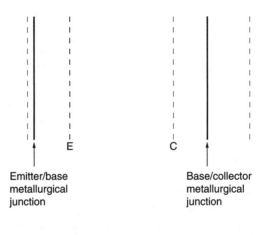

Emitter/base
metallurgical
junction

Base/collector
metallurgical
junction

Figure 7.11

The constant forward bias of the emitter–base junction causes an electron density at its transition region edge in the base which I shall call n_{p_1e}. From Equation (5.7) of Chapter 5

$$n_{p_1e} = n_{p0} \exp(eV_{be}/kT)$$

With zero volts between collector and emitter, the base–collector junction is also forward-biased by the same amount as the base–emitter junction, so that the electron density at its transition edge in the base, which I shall call n_{p_1c}, is the same as n_{p_1e}. Electrons diffuse away from both junctions towards the base contact, recombining with holes on the way. Holes also cross both junctions, from base to emitter and collector, respectively, and diffuse away, adding their currents to the electron current in each. The consequence is that, with zero collector–emitter volts, there is a current in the collector lead which is (in conventional current direction) out of the transistor – because of the relative dopings, this current is small. There is also a significant base current.

A small positive voltage applied between collector and emitter decreases the forward bias on the base–collector junction and reduces the value of the electron density at the collector depletion layer edge in the base to n_{p2c}, as shown

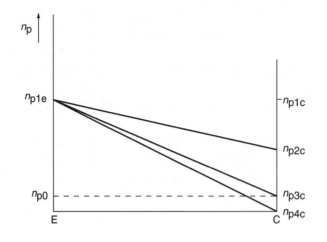

Figure 7.12

in Fig. 7.12. There is now a density gradient of electrons in the base which causes the electrons to diffuse to the collector and into its transition field, so, despite the fact that the base–collector junction is forward-biased, the electron flow is now from the base into the collector. There is still a flow of holes down a density gradient in the collector, but since the forward bias is reduced, this is reduced. You should see that, as the value of V_{CE} is increased, a point is quickly reached at which the electron current and hole current across the collector junction cancel so that the external value of I_C is zero.

As the collector voltage is further increased, the electron density gradient in the base increases, and the hole diffusion gradient in the collector decreases. Consequently the collector current increases, until the collector–base junction is reverse-biased by a small amount, when the electron density at the collector depletion layer edge has dropped to zero (shown as n_{p4c} in Fig. 7.12) and can drop no further. This will certainly be the case by the time the collector–emitter voltage has reached 1 V.

Self assessment test 7.1

What will be the collector–emitter voltage when the electron density at the collector transition region edge is n_{p3c} shown in Fig. 7.12?

Answer

The collector junction will have zero bias, so the collector–emitter voltage will be $+0.6$ V.

___

As the collector–emitter voltage is raised above about 1 V the collector current still increases, but very slowly. The reason for this increase is that, as the collector junction is increasingly reverse-biased, its transition region widens and the transition region edge marked C in Fig. 7.11 moves to the left, narrowing the part of the base between transition regions and increasing the density gradient. This is known as the 'Early' effect, named after the man who first reasoned it out.

Figure 7.13 illustrates the shape of typical output characteristic curves for different base voltages. If the scale were large enough I should have to show that the characteristics do not actually go through the origin, but meet the voltage axis at a very small positive voltage. For a fixed collector voltage, above about 1 V, the collector current is effectively the emitter junction current and so it changes with the base–emitter voltage exponentially, in the same way that a diode current changes with the diode-voltage.

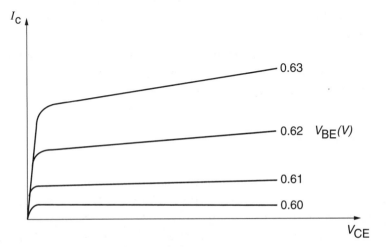

Figure 7.13   Bipolar transistor characteristics.

Other types of BJT

A pnp transistor will operate in a similar manner to that described for the npn device: holes are the minority carriers in the base, and the applied voltages and current directions are reversed. BJTs are also constructed in a number of different ways depending on the processing methods being used.

7.4   SUBSTRATES AND EPITAXIAL LAYERS

As we have seen, silicon planar devices and circuits are formed in a substrate: a flat piece of silicon, typically 200 μm thick, which has been manufactured with a designed doping density, n or p, with a known orientation of its crystal structure and within maximum specified impurity and dislocation density values. Devices can be made by introducing further dopants into the substrate by atomic diffusion or by ion implantation. This latter process is one whereby the required impurity atoms are ionized and 'fired' into the surface of the material electrically. The substrate, however, may not be sufficiently impurity- and dislocation-free, or it may not have the desired doping density; both of these are reasons for growing on it an epitaxial layer.

An epitaxial layer is a layer which is grown on a crystal so that it is continuous with the crystal structure; there is no discontinuity at the interface. Such layers are usually not more than 25 μm thick and can be grown slowly and carefully so that the crystal structure is as near perfect as possible. Doping can be included, as required, with the growing of the layer. The layer may be laid down from a gas or mixture of gases (vapour phase epitaxy) or it may be grown from a liquid in contact with the surface (liquid phase epitaxy). The whole of a device, or even the whole of a circuit, may be subsequently formed in the epitaxial layer, or the layer may form part of a device. Growing an epitaxial layer would be, for example, a way of producing an abrupt pn junction with uniform doping on either side.

It is common for a transistor which is part of an integrated circuit to be formed in an epitaxial layer on a main substrate which provides mechanical strength and may serve other functions.

7.5 SUMMARY

The ideal output characteristics for a transistor are shown in Fig. 7.2. The practical output characteristics of a MOSFET are represented in Fig. 7.9, and those of a bipolar junction transistor in Fig. 7.13.

A MOSFET has a structure as illustrated in Fig. 7.4. A voltage applied between the gate and source, when it reaches a threshold value, produces an inversion layer under the oxide, in which the minority and majority carrier densities are reversed compared to their values in the neutral substrate material. This provides a conducting channel between the source and drain regions. Voltage drop along the channel when a current flows causes the gate–channel voltage to be progressively lower from the source to the drain end, so that the channel carrier density, and hence the conductivity of the channel, drops progressively from the source to the drain end. When, for a given gate–source voltage, the drain–source voltage is sufficiently great that the gate–channel voltage drops below the threshold value at the drain end, the channel becomes pinched off with a high resistance section at the drain end. In this condition, changes in drain–source voltage are taken up by small changes in the length of the pinched-off section and the channel current remains almost constant.

The bipolar junction transistor consists essentially of two pn junctions formed close together and back-to-back. One particular structure is shown in Fig. 7.10. The emitter is heavily doped compared to the base, so that forward current in the emitter–base junction consists predominantly of carriers injected into the base from the emitter (the term 'injected' may seem inappropriate since the carriers diffuse across the transition region, but it is commonly used in the literature).

When the base–emitter junction is forward-biased and the collector–emitter voltage has the same polarity as this, carriers injected into the base diffuse down a density gradient to the collector and emerge as collector current. At low increasing collector voltages, the density gradient in the base increases rapidly, because the minority carrier density at the collector transition region edge is falling, and so the collector current increases rapidly. With increasing collector voltage a value is reached at which the collector–base junction is reverse-biased and the minority carrier density at the collector transition region edge becomes zero. Beyond this collector voltage the minority carrier density gradient in the base increases only slowly, and so the collector current increases slowly, because the collector transition region widens, narrowing the effective base width.

In silicon planar technology, transistors are often formed in a high quality epitaxial layer which has been grown on a rather less high quality substrate. The substrate provides mechanical strength and may have other functions in the operation of the device or circuit.

8 MOSFET parameters

In Chapter 7 I gave an outline of how MOSFETs function. Here I shall show how dopings and dimensions determine the various device parameters. I shall also deduce a small-signal equivalent circuit for the MOSFET.

8.1 CHARGE AT THE OXIDE–SILICON INTERFACE

Because of the difference in structure between crystalline silicon and silicon oxide, not all the valence bonds form to the surface silicon atoms; there are 'dangling bonds' which result in a positive charge on the interface. This positive charge is not mobile and consists, typically, of around 10^{15} electronic charges per square metre. I shall refer to it as the 'oxide surface charge'.

It is possible to modify the effective surface charge density by adding a surface layer of acceptor or donor atoms in the silicon. Consider, for example, the n-channel device discussed in Chapter 7. The oxide surface charge will cause an n channel to start to form; if a channel forms without a positive voltage on the gate the device is depletion mode. A thin extra layer of acceptor atoms will release a few more holes into the substrate and form a surface layer of fixed negative charge which will counteract the oxide layer and ensure that the device is enhancement mode. On the other hand, a surface layer of donors will release a few electrons and add extra fixed positive charge to ensure that the device is definitely depletion mode. Surface doping will similarly modify a p-channel device, so that enhancement mode and depletion mode versions of both channel polarities can be made.

8.2 THRESHOLD VOLTAGE

I shall start, again, with an n-channel MOSFET, which I shall assume is enhancement mode despite the oxide surface charge. Assume that the gate–substrate voltage is V_T, and that a positive charge is on the gate. Look at Fig. 8.1, which is a diagram of the charge distribution in a cross-section through the gate and substrate. I am going to treat the depleted region as having

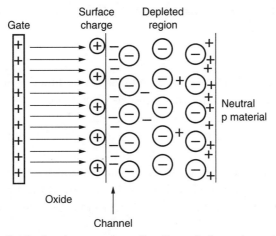

Gate Surface charge Depleted region

Oxide

Neutral p material

Channel

Figure 8.1 Charge distribution in a cross-section through the gate and substrate of an n-channel MOSFET.

abrupt edges; and in addition to the fixed acceptor ions contained therein, there are the electrons under the oxide which form the channel and an equal number of holes near the neutral region edge which provide a hole diffusion gradient.

The total voltage between gate and substrate is the sum of the voltage across the oxide and the voltage across the depleted region. Let us start with the second of these. I explained in Chapter 7 that this voltage is related to the mobile charge distribution by the equation

$$V_{cs} = (kT/e) \ln(n_c/n_s)$$

At the threshold voltage, $n_c = N_A$ and $n_s = n_{p0} = n_i^2/N_A$, so the voltage across the depleted region, represented for this particular condition by ψ_D, is given by

$$\psi_D = (kT/e) \ln(N_A^2/n_i^2)$$

which can be rewritten

$$\psi_D = \frac{2kT}{e} \ln\left(\frac{N_A}{n_i}\right) \tag{8.1}$$

Now I must consider the voltage across the oxide. The oxide forms the dielectric of a capacitor of which the gate electrode forms one 'plate' and the depleted region the other. Looking at Fig. 8.1 and applying Gauss's theorem to the depleted region, the net electric flux in the oxide corresponds to the net negative charge in the depleted region, which, since there is an equal number of hole as electrons in the region (although in different parts), has a value

$$Q_D - Q_{ss}$$

where Q_{SS} is the net surface charge, in this case simply the oxide surface charge, and

$$Q_D = eN_A Aw \qquad (8.2)$$

where A is the cross-sectional area and w is the depleted region width.

So now I have to calculate the depleted region width. I can make the simplifying assumption that the total charge in the inversion layer and in the hole density gradient within the depleted region is negligible compared to the total acceptor ion charge in the depleted region. I can then use Gauss's theorem to obtain electric field as a function of distance and integrate this to obtain voltage as I did in Chapter 5 when I found the width of the transition region in a pn junction. The result in this case, similar to that for one side of a pn junction, is

$$\psi_D = (eN_A/\epsilon_0\epsilon_r)w^2/2$$

ψ_D being the voltage across the depleted region for this condition. Rearranging gives

$$w = \sqrt{\left(\frac{2\epsilon_0\epsilon_r\psi_D}{eN_A}\right)} \qquad (8.3)$$

Substituting Equation (8.3) into (8.2) gives

$$Q_D = A\sqrt{(2\epsilon_0\epsilon_r\psi_D eN_A)} \qquad (8.4)$$

The oxide layer capacitance, which can conveniently be defined per unit area, is given by

$$C_{ox} = \epsilon_0\epsilon_{ox}/t_{ox}$$

where t_{ox} is the oxide thickness and ϵ_{ox} is the relative permittivity of silicon oxide ($=4$). So, the voltage across the oxide is

$$\frac{Q_D - Q_{SS}}{AC_{ox}}$$

Finally then, for this particular case,

$$V_T = \frac{Q_D - Q_{SS}}{AC_{ox}} + \psi_D \qquad (8.5)$$

If surface doping had been added, then Q_{SS} would be the resultant of the oxide surface charge and the added surface charge. I can modify Equation (8.5) to apply to any MOSFET as follows:

$$V_T = \frac{\left|\dfrac{Q_D}{A}\right| + \left|\dfrac{Q_{SS}}{A}\right|}{C_{ox}} \pm |\psi_D| \qquad (8.6)$$

In Equation (8.6), $|Q_D|$, $|Q_{SS}|$ and $|\psi_D|$ are positive magnitudes irrespective of the signs or polarities of the quantities. If the sign of Q_{SS} is the same as that of Q_D then the $\pm$ is a plus; if it is opposite then it is a minus. If V_T comes out positive, then the device is enhancement mode; the actual polarity of the gate–substrate voltage should be clear from the channel type. In the case where Q_{SS} is greater than Q_D and opposite in sign, V_T can come out as a minus quantity; this indicates that the device is depletion mode and actually requires a 'reverse' gate voltage to turn it off.

In Equation (8.6)

$$\psi_D = \frac{2kT}{e}\ln\left(\frac{N_{A/D}}{n_i}\right) \tag{8.7}$$

$$\frac{Q_D}{A} = \pm\sqrt{(2\epsilon_0\epsilon_r e\psi_D N_{A/D})} \tag{8.8}$$

$$\frac{Q_{SS}}{A} = e(N_{ox} \pm N_{sd}) \tag{8.9}$$

and

$$C_{ox} = \frac{4\epsilon_0}{t_{ox}} \tag{8.10}$$

$N_{A/D}$ is the doping density in the substrate, whichever polarity, N_{ox} is the number of oxide surface charges per unit area (always positive charge) and N_{sd} is the number of surface doping charges per unit area (which can be positive or negative charge).

If the gate is made of metal, then a further voltage must be added to V_T in Equation (8.6) to take account of the difference in work function between the gate material and the substrate material. I have not included this because in modern devices, particularly in integrated circuits, the gate is made of doped polycrystalline silicon which acts as a good conductor and has the same work function as the substrate.

Self assessment test 8.1

A p-channel MOSFET with a polysilicon gate has a substrate doping density of 5×10^{22} donors/m³. The oxide is 10^{-7} m thick and has 10^{15} surface charges/m². The device is at room temperature.
(a) Assuming that the source is connected to the substrate, deduce the required operating polarities of the gate–source and drain–source voltages.
(b) Show that with no added surface doping the device is enhancement mode and calculate the threshold voltage.
(c) Find the minimum surface doping and type of dopant required to make the device depletion mode.

Answers

(a) To form a p channel in an n-type substrate, the gate–source voltage must be negative. The drain–source voltage must also be negative so that the drain region is isolated from the substrate.

(b) Substituting into Equation (8.7), since $kT/e = 0.025$ V at room temperature,

$$\psi_D = 2 \times 0.025 \ln(5 \times 10^{22}/10^{16})$$

$$= 0.77 \text{ V}$$

In Equation (8.8)

$$Q_D/A = \sqrt{(2 \times 8.85 \times 10^{-12} \times 12 \times 1.6 \times 10^{-19} \times 0.77 \times 5 \times 10^{22})}$$

$$= 1.14 \times 10^{-3} \text{ C m}^{-2}$$

This is positive because the dopant ions are donors.

In Equation (8.9)

$$Q_{SS}/A = 1.6 \times 10^{-19} \times 10^{15}$$

$$= 1.6 \times 10^{-4} \text{ C m}^{-2}$$

In Equation (8.10)

$$C_{ox} = (4 \times 8.85 \times 10^{-12})/10^{-7}$$

$$= 3.54 \times 10^{-4} \text{ F m}^{-2}$$

In Equation (8.6)

$$V_T = (1.14 \times 10^{-3} + 1.6 \times 10^{-4})/3.54 \times 10^{-4} + 0.77$$

$$= 4.44 \text{ V}$$

Since this answer comes out positive the device is enhancement mode. The actual threshold requires the gate to be negative to the source by 4.4 V.

(c) Surface doping will only alter the value of Q_{SS}. From Equation (8.6) one sees that negative surface doping is required, sufficient to cancel the effect of the oxide surface charge, the donor charge in the depleted region and of ψ_D. To make V_T zero requires

$$|Q_{SS}|/A = |Q_D|/A + |\psi_D|C_{ox}$$

$$|Q_{SS}|/A = 1.14 \times 10^{-3} + 0.77 \times 3.54 \times 10^{-4}$$

$$= 1.41 \times 10^{-3} \text{ C m}^{-2}$$

Adding to this the oxide surface charge to be neutralized, the total dopant surface charge required is

$$1.41 \times 10^{-3} + 1.6 \times 10^{-4} = 1.57 \times 10^{-3} \text{ C m}^{-2}$$

The dopant required is an acceptor, to make the surface negative, and the surface density required to make V_T zero is

$1.56 \times 10^{-3}/1.6 \times 10^{-19} \approx 10^{16}$ atoms m^{-2}

More doping than this will make the device depletion mode: a little more calculation shows that 2×10^{16} acceptors/m^3 will give a threshold gate–source voltage of about +4.5 V.

8.3 EQUATIONS FOR THE OUTPUT CHARACTERISTICS

Figure 8.2 shows the channel of an n-channel enhancement mode MOSFET with a current flowing. I am considering an n-channel device again because all the appropriate applied voltages are positive. I shall make the simplifying assumptions that the channel has a definite constant depth, d, and that variations in the voltage between the gate and the channel, above the threshold value, cause proportional variations to the carrier density in the channel. The channel width is W, as shown.

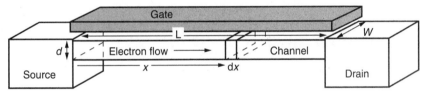

Figure 8.2

Consider a small length of the channel, dx, at a distance x from the source. The voltage in the channel at this point, relative to the source, is V_x. The magnitude of the voltage which is inducing carriers at this point is

$V_{GS} - V_x - V_T$

The amount of charge, in the form of carriers, induced by this voltage in the length of channel dx is (remembering that C_{ox} is capacitance per unit area)

$Q_{dx} = C_{ox}Wdx(V_{GS} - V_x - V_T)$

If n is the number of carriers per unit volume, then

$n = Q_{dx}/eWddx = C_{ox}(V_{GS} - V_x - V_T)/ed$

from which

$$ned = C_{\text{ox}}(V_{\text{GS}} - V_x - V_{\text{T}}) \tag{8.11}$$

Since the current is entirely drift, the magnitude of the current density is given by

$$J = en\mu_n E_x$$

and so the total channel current magnitude (same as the drain current) is

$$I_{\text{D}} = Wden\mu_n E_x$$

I can substitute dV_x/dx for E_x (you will see that I am dealing throughout with magnitudes and ignoring signs), and the expression for *ned* in Equation (8.11), to give

$$I_{\text{D}} = W\mu_n C_{\text{ox}}(V_{\text{GS}} - V_x - V_{\text{T}})(dV_x/d_x)$$

Separating variables to integrate along the length of the channel gives

$$\int_0^L \frac{I_{\text{D}}}{W\mu_n C_{\text{ox}}}\, dx = \int_0^{V_{\text{DS}}}(V_{\text{GS}} - V_x - V_{\text{T}})dV_x$$

which yields

$$\frac{I_{\text{D}}L}{W\mu_n C_{\text{ox}}} = (V_{\text{GS}} - V_{\text{T}})\,V_{\text{DS}} - \tfrac{1}{2}V_{\text{DS}}^2$$

So

$$I_{\text{D}} = \beta V_{\text{DS}}[(V_{\text{GS}} - V_{\text{T}}) - \tfrac{1}{2}V_{\text{DS}}] \tag{8.12}$$

where

$$\beta = \frac{W\mu_n C_{\text{ox}}}{L} \tag{8.13}$$

β is called the 'gain factor': its SI units are amperes per volt squared (AV^{-2}). As you can see, the value of β depends entirely on the structure of the device. To calculate β, a value for μ_n about half that given in Table 1.2 must be used because defects near the surface cause the mobility there to be lower than in the bulk of the material.

[***Note*** Note that the symbol β is used to represent an entirely different parameter for bipolar junction transistors.]

Equation (8.12) holds so long as the channel is not pinched off. Suppose that at the drain ($V_{\text{GS}} - V_x - V_{\text{T}}$) just equals zero. V_x at the drain is V_{DS}, so for this condition,

$$V_{\text{DS}} = V_{\text{GS}} - V_{\text{T}}$$

and from Equation (8.12)

$$I_D = \beta V_{DS}(V_{DS} - \tfrac{1}{2}V_{DS}) = \tfrac{1}{2}\beta V^2_{DS}$$

or

$$I_D = \tfrac{1}{2}\beta(V_{GS} - V_T)^2 \tag{8.14}$$

When V_{DS} is increased beyond this critical value a short length of the channel is 'pinched off' with a high resistance. This is equivalent to shortening the channel, so in Equation (8.14), β must be replaced by the expression

$$\frac{W\mu_n C_{ox}}{L(1 - \lambda V_{EX})} \tag{8.15}$$

The parameter λ is called the 'channel length modulation factor'; V_{EX} is the excess drain–source voltage above that required just to pinch off the channel, so

$$V_{EX} = V_{DS} - (V_{GS} - V_T)$$

Provided $\lambda V_{EX} \ll 1$, which it generally is, the expression (8.15) can be rewritten

$$\frac{W\mu_n C_{ox}}{L}(1 + \lambda V_{EX})$$

that is,

$$\beta[1 + \lambda(V_{DS} + V_T - V_{GS})]$$

So Equation (8.14) becomes

$$I_D = \tfrac{1}{2}\beta(V_{GS} - V_T)^2[1 + \lambda(V_{DS} + V_T - V_{GS})] \tag{8.16}$$

Equations (8.12) and (8.16) together, and (8.13) for the value of β, give the characteristic curves for any MOSFET (you need to think about the appropriate polarities of the voltages).

If I put $I_D = 0$ in Equation (8.16), it follows that

$$1 + \lambda(V_{DS} + V_T - V_{GS}) = 0$$

$$V_{DS} = (V_{GS} - V_T) - \frac{1}{\lambda}$$

The value of $1/\lambda$ is generally at least 50 V, while $(V_{GS} - V_T)$ is only a few volts, so to a good approximation, when $I_D = 0$, $V_{DS} = -1/\lambda$. The consequence of this is illustrated in Fig. 8.3; the characteristics beyond pinch-off appear to diverge from a point on the voltage axis at about $-1/\lambda$ volts.

The previous discussion suggests, and it proves to be the case in practice, that the pinched-off length for a given value of V_{EX} is effectively the same, irrespective of the channel length. $L(1 - \lambda V_{EX})$ in the denominator of expression (8.15) is the same as $L - L\lambda V_{EX}$, indicating that the reduction in

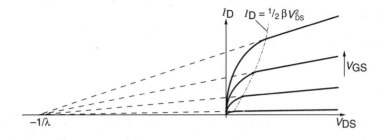

Figure 8.3   n-channel MOSFET characteristics.

length is $L\lambda V_{EX}$; so since this is found to be independent of L, it follows that λ must be inversely proportional to L, i.e. a long channel yields a small value of λ.

8.4   TRANSCONDUCTANCE AND DRAIN SLOPE CONDUCTANCE

The required property of the MOSFET is that the gate voltage should control the drain current. The rate of change of drain current with gate–source voltage, dI_D/dV_{GS}, for fixed drain–source voltage, is the small-signal mutual conductance or transconductance, g_m.

For values of V_{DS} below pinch-off, differentiating Equation (8.12) yields

$$g_m = \beta V_{DS}$$

so under these conditions, for a fixed V_{DS}, the transconductance is independent of V_{GS}. For this reason this part of the characteristics is called the 'linear region' even though the characteristic curves are anything but linear!

The region of operation beyond pinch-off is known as the 'saturated region'. To find dI_D/dV_{GS} for fixed V_{DS} accurately from Equation (8.16) is very difficult; however, assuming that λ is very small, an approximate result is

$$g_m = \sqrt{(2\beta I_D)}$$

Here the transconductance is proportional to the square root of the drain current.

A large value of g_m is necessary for high amplification in analogue circuits. You will see from both formulae above that the value of β determines the range of g_m, hence the name 'gain factor'.

The slope of the output characteristic, dI_D/dV_{DS}, for fixed V_{GS} at any point is called the drain slope conductance at that point. Taking first the linear region, differentiating Equation (8.12) with respect to V_{DS} yields

$$g_{ds} = \beta(V_{GS} - V_T - V_{DS}) \tag{8.17}$$

In the saturated region, differentiating Equation (8.16) is again difficult, but from Fig. 8.3 I can deduce that, to a good approximation,

$$g_{ds} = \frac{I_D}{V_{DS} + 1/\lambda}$$

To have as small a value of g_{ds} as possible is generally desirable in analogue circuits. Apart from putting values in Equations (8.17) and (8.18), you can see from Fig. 8.3 that the slope of the characteristics is much smaller in the saturated than in the linear region, so in analogue circuits MOSFETS are generally used in the saturated region. Smaller drain currents levels also give lower g_{ds} values.

For a small value of g_{ds} in the saturated region, a large value of $1/\lambda$ is required. This indicates the need for a long channel.

8.5   THE SMALL-SIGNAL EQUIVALENT CIRCUIT

The MOSFET is not a linear device and cannot be represented by a linear equivalent circuit; however, when one is used in a small-signal analogue circuit it is expedient for the purposes of analysis to represent it by a small-signal (quasi-linear) equivalent circuit. The equivalent circuit which works for a MOSFET is shown in Fig. 8.4 and I shall relate its parameters to the physical properties of the device.

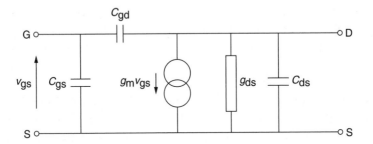

Figure 8.4   A MOSFET equivalent circuit.

Comparing Fig. 8.4 with Fig. 7.3 in Chapter 7, which represents the ideal, you will see that we would like all three capacitances and g_{ds} to be as small as possible. g_m and g_{ds} have been discussed in previous sections.

C_{gs} and C_{gd} both derive from the capacitance between the gate and the channel. This capacitance is essential to the operation of the device. Its size depends on the area of the gate and on the oxide thickness. From Equation (8.6) you will see that for a given required threshold voltage a smaller value of C_{ox} indicates lower substrate doping and a careful control of surface charge. To a first approximation C_{gs} and C_{gd} can be assumed to each be equal to half the gate–channel capacitance.

Remembering that the source is connected to the substrate, the drain is isolated from the substrate by a reverse-biased pn junction and this involves a transition region capacitance. This capacitance is the main component of C_{ds}.

8.6   MOSFETS IN DIGITAL INTEGRATED CIRCUITS

MOSFETs are used widely in digital integrated circuits. Two 'families' of such integrated circuits are known as 'NMOS' and 'CMOS'. In the NMOS family all the devices are n-channel; CMOS uses both n-channel and p-channel devices, the 'C' standing for 'complementary'. The digital circuits concerned are all based on switching, and in the case of both technologies, the switching transistor is an enhancement mode device which will be definitely 'on' (conducting) when it has a voltage on its gate above the threshold, and definitely 'off' when it has zero volts on its gate. It is not my purpose in this book to consider circuitry (there are many excellent texts on digital circuit design); I simply want to draw your attention to one or two structural issues, and to do so I shall consider, as an example, an inverter in each family.

NMOS

In an NMOS inverter, besides the switch, another transistor, usually depletion mode, acts as a load. It is important that the switching transistor has as high a gain as possible; this determines how quickly it switches. Hence the value of β for the switching transistor should be as high as possible. On the other hand, again to keep the gain high, the load transistor needs to have as high a slope resistance (hence as low a value of g_{ds}) as possible.

Looking at the expression for β in Equation (8.13) you will see that to make β large, W needs to be large and L small; in the switching transistor the width of the channel might well be twice its length. The load transistor, on the other hand, needs a high value of L to give it a large g_{ds}. For practical construction reasons both transistors will be made with similar widths, but in the case of the load transistor the length might typically be twice the width. So the two transistors, side by side on the chip, have very different channel lengths.

Another significant practical consideration is that the two devices are connected in series but share a common substrate, so that only one source, the switching transistor source, can be connected to the substrate. The source of the load transistor is connected to the drain of the switch; in fact the two are generally the same n$^+$ region. This causes the load transistor threshold voltage to change when the switching transistor switches between 'on' and 'off'. The practical result of this is that the effective drain slope resistance is reduced.

Figure 8.5 shows an NMOS inverter in cross-section.

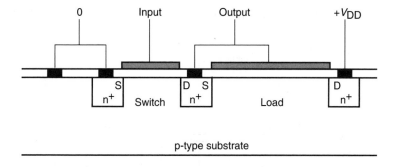

Figure 8.5   Structure of an NMOS inverter.

Self assessment test 8.2

Deduce from Fig. 8.5 why, with respect to the zero line, the output is near zero when the input is set to the positive line voltage $(+V_{DD})$, while with the input set to zero the output is $+V_{DD}$.

Explanation

The load depletion mode transistor is permanently conducting since its gate and source are connected together. With $+V_{DD}$ on the input terminal the switch transistor is conducting, and, since it has a lower resistance than the load most of the positive line voltage is dropped in the load, setting the output voltage near to zero. With the input terminal set to zero the switch transistor is not conducting, so the output is at $+V_{DD}$.

CMOS

In a CMOS inverter an n-channel enhancement mode transistor and a p-channel enhancement mode transistor are connected in series, and the drive is connected so that when the input is high the n-channel device is on while the p-channel device is off, and when the input is low the p-channel device is on while the n-channel device is off. Current only flows through the two while switching is taking place, so the power consumption is very low. Both transistors act as switches, so both need gains that are as high as possible. β for the p-channel device will be given by Equation (8.13), modified by replacing μ_n by μ_p. Since the value of μ_p for silicon is less than one-third of the value of μ_n, the switching times of the two transistors will not be the same, unless the channel length of the p-channel device is considerably shorter than that of the n-type. Generally both channel lengths are made as short as is practicable, and the slower switching of the p-channel device is accepted. Figure 8.6 shows a CMOS inverter in cross-section. Notice that to accommodate the p-channel transistor a 'well' of n-type doping has been introduced into the p-type substrate.

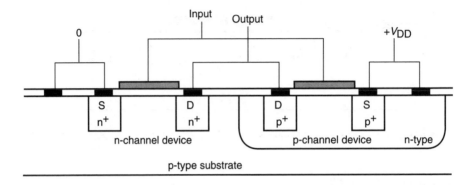

Figure 8.6   Structure of a CMOS inverter.

8.7   SUMMARY

The threshold voltage of a MOSFET can be found from the equation

$$V_T = \frac{\left|\dfrac{Q_D}{A}\right| \pm \left|\dfrac{Q_{SS}}{A}\right|}{C_{ox}} + |\psi_D|$$

where ψ_D is the voltage drop across the depleted region and is given by

$$\psi_D = \frac{2kT}{e} \ln\left(\frac{N_{A/D}}{n_i}\right)$$

Q_D/A is the ionic charge per unit gate area in the depleted region, given by

$$\frac{Q_D}{A} = \pm\sqrt{(2\epsilon_0\epsilon_r e\psi_D N_{A/D})}$$

Q_{SS}/A is the net surface charge, the resultant of oxide surface charge and surface doping, per unit gate area and is given by

$$\frac{Q_{SS}}{A} = e(N_{ox} \pm N_{sd})$$

C_{ox} is the capacitance (of which the oxide is the dielectric) per unit gate area, and is given by

$$C_{ox} = \frac{4\epsilon_0}{t_{ox}}$$

and $N_{A/D}$ is the doping density in the substrate. The plus or minus in the numerator of the fractional expression, for V_T is plus if Q_D and Q_{SS} have the same polarity and minus if they are opposite.

If V_T comes out positive the device is enhancement mode; if negative it is depletion mode.

A MOSFET is operating in the linear region if

$$V_{DS} < V_{GS} - V_T$$

and in the saturated region if

$$V_{DS} > V_{GS} - V_T$$

The output characteristics of a MOSFET are represented in the linear region by the equation

$$I_D = \beta V_{DS}[(V_{GS} - V_T) - \tfrac{1}{2}V_{DS}]$$

and in the saturated region by

$$I_D = \tfrac{1}{2}\beta(V_{GS} - V_T)^2[1 + \lambda(V_{DS} + V_T - V_{GS})]$$

where

$$\beta = \frac{W\mu C_{ox}}{L}$$

μ being half the value of that quoted in Table 1.2 for the appropriate carriers.

In analogue circuits, MOSFETS are generally used in the saturated region of operation where

$$g_m \approx \sqrt{(2\beta I_D)}$$

and

$$g_{ds} \approx \frac{I_D}{V_{DS} + 1/\lambda}$$

An analogue small-signal equivalent circuit for a MOSFET is shown in Fig. 8.4.

In NMOS digital switching circuits enhancement mode n-channel transistors with short channels are used as switches, and n-channel depletion mode transistors with longer channels are used as loads.

CMOS switching circuits use a 'push–pull' arrangement with two enhancement mode transistors, one n-channel and one p-channel. In each state of the switch, one transistor is conducting and the other is not, so that current only flows through the two during switchover.

8.8 PROBLEMS

A MOSFET is formed in a silicon substrate doped with boron atoms to a density of 10^{23} m^{-3}. The oxide layer, under a p-type polysilicon gate, has a thickness of 0.1 μm. Under the oxide is a layer of phosphorus atoms with a surface density of 1.5×10^{16} m^{-2}. The oxide surface charge can be taken to be $10^{15}e$ m^{-2}. The width of the channel is $15\,\mu$m and its length is $10\,\mu$m. λ is 0.01 V^{-1}.

1 Calculate the threshold voltage and deduce what type the transistor is.

2 Find the drain current when the drain–source voltage is $+10$ V and the gate–source voltage is $+5$ V.

3 For the operating point given in question 2, find the approximate values of g_m and g_{ds}.

4 Estimate the values of C_{gd} and C_{gs} in the equivalent circuit.

9 Bipolar junction transistor parameters

This chapter develops the ideas introduced in Chapter 7 so as to explain in somewhat more detail the properties of BJTs. I shall continue to relate my explanations to the silicon planar form of construction.

9.1 RESISTIVE BASE CURRENT

It should be clear from the account given in Chapter 7 that there is an unavoidable capacitive current in the base lead of a BJT whenever the base or collector voltage changes – unavoidable because the charge distribution in the base and in the transition regions changes. However, in an ideal device, if the base and collector voltages were fixed, with the base–emitter junction forward-biased and the emitter–collector junction reverse-biased, there would be no current in the base lead. In practice there is, for the following reasons.

The first reason that resistive base current flows is that the emitter efficiency is not 100 per cent. The current crossing the emitter–base junction is predominantly made up of majority carriers from the emitter injected into the base; to ensure that this is so, the emitter is much more heavily doped than the base. However, a small proportion of the emitter–base current consists of majority carriers from the base entering the emitter. This fraction of current plays no part in the transistor action, and accounts for most of the current in the base lead. The ratio of the injection current into the base to the injection current into the emitter can be calculated from Equation (5.12) of Chapter 5. From this the emitter efficiency, which is the fraction of the total emitter current represented by injection into the base, can be deduced.

A second cause of base current is recombination. In any practical transistor the active base width will be much smaller than the diffusion length of minority carriers, so it is reasonable to treat the density gradient as linear, implying no recombination in the base. However, the notion of diffusion length, and that of minority carrier lifetime from which it is derived, are statistical ones; there will always be some minority carriers which recombine with majority carriers quite soon after entering the base. One must add to this a certain amount of recombination in the emitter–base junction and recombination near the base connection of carriers which have diffused out of the active region. There will be a small flow of base current to replenish the majority carriers lost in recombination.

Finally, there is the collector junction leakage current. With the collector–base junction reverse-biased, the part of its normal reverse saturation current which consists of minority carriers from the collector injected into the base results in a third component of base current. This contribution is, however, generally very much smaller than the other two.

The collector current and base current both flow in the same direction – either out of the transistor for a pnp device or into the transistor (conventional current direction) for an npn device. Kirchhoff's first law indicates that

$$I_C + I_B = I_E$$

The ratio I_C/I_E is usually given the symbol α. More important is the ratio I_C/I_B, symbol β, which is the 'current gain' of the transistor.

[**Note**   Strictly, α is the 'common-base current gain'. It is the current 'gain' (actually a loss) between the emitter circuit and collector circuit if the base is made the common electrode. Thus the full description of β is the 'common-emitter current gain'.]

Self assessment test 9.1 _____

Deduce the relationship between α and β.

Answer

$$\beta = \frac{I_C}{I_B} = \frac{I_C}{I_E - I_C} = \frac{\alpha}{1 - \alpha}$$

Generally, α is very nearly 1 and I_B is very small, so where necessary I can take $I_C \approx I_E$ and $\beta \approx I_E/I_B$.

9.2   INPUT RESISTANCE

The voltage between base and emitter is the voltage across a forward-biased pn junction, and the current into the base is $1/\beta$ of the pn junction current, so the input resistance to the transistor, between the base and emitter terminals, is β times that of the diode equivalent of the emitter–base junction. A diode has a non-linear resistance, so the input resistance to a BJT is also non-linear.

For small signals, we have seen in Chapter 6 that the slope resistance of the diode is

$$r_e = kT/eI_D$$

so for the transistor, the small-signal input resistance is

$$r_i = \beta r_e$$

where r_e is now defined as kT/eI_E.

At room temperature $r_e = 25/(I_E$ in mA). Suppose that $\beta = 200$ (a reasonable value); with a collector current of 1 mA (and an emitter current effectively the same) this indicates a small-signal input resistance of 5000 Ω. For larger direct collector currents the small-signal input resistance will be proportionately smaller.

9.3   TRANSCONDUCTANCE

The relationship between the collector current and the base–emitter voltage is approximately the same as that between the emitter current and the base–emitter voltage, so I can write, as for a diode

$$I_C \approx I_s \exp(eV_{BE}/kT)$$

Again, the relationship between I_C and V_{BE} is not linear, so the value of I_C/V_{BE} has little meaning, but the small-signal transconductance, defined as dI_C/dV_{BE}, is given by

$$g_m = \frac{eI_C}{kT} \approx \frac{1}{r_e}$$

This indicates that the small-signal transconductance should be the same for all bipolar junction transistors, depending only on the collector current. At room temperature

$$\frac{1}{re} = \frac{I_C(\text{in mA})}{25} \text{ A V}^{-1} = \frac{1000}{25} I_C(\text{in mA}) \text{ mA V}^{-1}$$

$$g_m = 40 \, I_C(\text{in mA}) \text{ mA V}^{-1}$$

9.4   OUTPUT CHARACTERISTICS AND SLOPE CONDUCTANCE

Look at Fig. 9.1 which shows the characteristics of an npn transistor. The knee of each curve occurs where the base–collector junction is reverse-biased by a

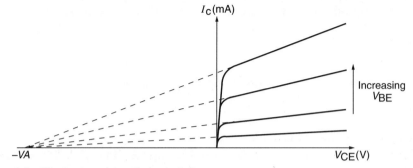

Figure 9.1   Bipolar transistor characteristics.

fraction of a volt. Since the values of V_{BE} for the normal range of collector currents lie between 0.6 and 0.7 V, that indicates that all the curves have their knees at around 1 V. The part of the characteristics for V_{CE} less than 1 V is called the 'saturation region', for collector voltages above about 1 V the transistor is in the 'active region'.

[***Note***   It is irritating that similar terms (in this case 'saturation region') are used for functionally different properties of BJTs and MOSFETs, but that is how the subject has evolved.

You will have noticed that the term 'active' is applied, in the context of BJTs, to the region of the device in which the transistor action takes place, to the base width between depletion layer edges and to the range of bias conditions for which the characteristics are beyond the 'knee'; there seems to be a shortage of appropriate adjectives!]

In the active region the characteristics slope because of the Early effect mentioned in Chapter 7. The Early effect results in 'base width modulation' in which the active width of the base decreases as the collector voltage is increased. This is similar in effect to channel length modulation in a MOSFET and has a similar result – all the characteristics, above the knees, appear to diverge from a point on the negative V_{CE} axis. For any BJT the magnitude of this voltage is called the 'Early voltage', symbol VA. (A pnp transistor uses negative V_{CE} values and its Early voltage is positive.)

The slope conductance of a characteristic, in the active region, is given the symbol g_o and

$$g_o = \frac{I_C}{V_{CE} + VA}$$

The Early effect may also dictate the maximum collector voltage that can be applied. If, as the value of V_{CE} is increased, the electric field in the collector–base junction reaches a value at which avalanche breakdown starts to occur, then this voltage must not be exceeded, but if no avalanching occurs, then the limit is set where the active base width is reduced to zero – a condition which is known as 'punch-through'.

9.5 THE SMALL-SIGNAL EQUIVALENT CIRCUIT

An equivalent circuit which gives a good representation of the behaviour of an appropriately biased planar bipolar junction transistor for small signals over a wide range of frequencies is shown in Fig. 9.2. The actual base terminal is at B, but there is a notional internal point, marked B' in the figure, which is the input to the active part of the base. Between B and B' is an ohmic resistance, $r_{bb'}$ which is due to bulk resistance of the base material between the base terminal and the active part of the base. Because of the layout and the fact that the base must be relatively lightly doped, this resistance is significant and typically has a value $\approx 30\ \Omega$. The bulk resistance of the emitter material is negligible because the emitter is heavily doped. The bulk resistance of the collector can be made small by making the collector material heavily doped except near the base–collector junction; for example, by forming the transistor in a lightly doped n-type epitaxial layer on a heavily doped n-type substrate.

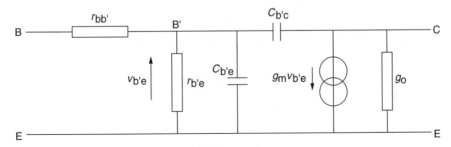

Figure 9.2 An equivalent circuit for a bipolar transistor.

$r_{b'e}$ is the input resistance, and is equal to βr_e. g_m is the transconductance, already discussed, and g_o is the output characteristic slope conductance. The capacitance $C_{b'e}$ is the sum of the emitter–base transition region capacitance and the diffusion capacitance of the excess minority carriers in the base. $C_{b'c}$ represents the transition region capacitance of the base–collector junction. The values of all the components mentioned in this paragraph depend on the 'd.c. bias' conditions, that is, on the direct voltages and currents applied.

Variation of β with collector current and its effect on $r_{b'e}$

The value of $r_{b'e}$ must vary with I_C because r_e varies with I_C; however, it turns out that the value of β also varies with I_C which causes a further variation in $r_{b'e}$.

The reason for the variation of β with I_C is as follows. As the collector current increases, the diffusion gradient of minority carriers in the base increases and so does that of majority carriers. There has to be a difference between the two – enough to leave a residual field to oppose the diffusion of the majority carriers. At low currents this residual field is small, but as the current is increased and hence the build-up of minority carriers increases, the residual field has the effect of imposing a component of drift on the minority carrier current. Consequently the transit time of the minority carriers across the base decreases and so recombination is reduced. Ultimately β approaches a value that depends on the emitter efficiency alone.

Because of the way they are constructed, planar bipolar junction transistors have a doping gradient in the base, such that the emitter end is more heavily doped than the collector end. This results in an in-built electric field which again imposes some drift on the minority carriers and tends to raise the low-current value of β.

In a typical planar transistor the value of β for the highest currents for which the transistor can be used is about twice its value for the lowest currents.

Typical equivalent circuit component values

Figure 9.3 shows an npn silicon planar transistor with dimensions and doping densities appropriate for a device which may form part of an integrated circuit. I shall assume that the area of the active part of the collector junction is the same as that of the emitter junction, although in practice there will probably be some current spreading, so that the active area is slightly larger at the collector than at the emitter. I am also assuming, for simplicity, that the doping is uniform and the junctions abrupt.

I have to choose a typical value for the d.c. base–emitter voltage, V_{BE}; I shall take it to be 0.60 V. This will fix the emitter–base transition region width, which I shall calculate first. However, before I can do this I need the values of the two junction contact potentials.

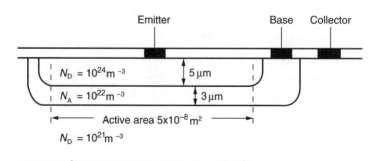

Figure 9.3

Self assessment test 9.2 _____

Using Equation (5.5) of Chapter 5,

$$\psi = (kT/e)\ln[n_i^2/(N_A N_D)]$$

find the contact potentials for the two junctions.

Answers

For the emitter–base junction $\psi = -0.81$ V; for the base–collector junction $\psi = -0.69$ V.

_____

Using Equation (5.15) of Chapter 5,

$$w_t = \sqrt{\left[\frac{2\epsilon_0 \epsilon_r}{e}\left(\frac{1}{N_A} + \frac{1}{N_D}\right)(|\psi| - V_D)\right]}$$

the width of the emitter–base junction transition region works out to

$$w_{teb} \approx 0.17 \mu m$$

Since the emitter is 100 times as heavily doped as the base, effectively all this transition region is in the base.

Let us assume now that the collector–emitter voltage, V_{CE}, is 2 V. This makes the base–collector voltage equal to -1.4 V. The base–collector transition region width works out to

$$w_{tbc} = 1.75 \ \mu m$$

In this case, since the base is 10 times as heavily doped as the collector, only 1/11 of the transition region is in the base. The base width reduction due to the base–collector transition region is therefore 0.16 μm.

The 'active' base width, which I shall call W, is 3 μm − 0.17 μm − 0.16 μm; therefore

$$W = 2.67 \ \mu m$$

The collector current can now be calculated. Assuming negligible recombination in the base, the electron density gradient is n_{p1}/W, where n_{p1} is the electron density at the emitter end of the active base region. Equation (5.7) can be applied here as

$$n_{p1} = n_{p0}\exp(eV_{BE}/kT)$$

and, since $n_{p0} = n_i^2/N_a$

$$n_{p1} = (10^{32}/10^{22})\exp(0.60/0.025)$$

$$n_{p1} = 2.6 \times 10^{20} \ m^{-3}$$

The collector current, which is the electron current through the base, is

$$I_C = AeD_n dn/dx$$

$$= 5 \times 10^{-8} \times 1.6 \times 10^{-19} \times 0.0039 \times 2.6 \times 10^{20}/2.67 \times 10^{-6}$$

$$I_C = 3.0 \text{ mA}$$

Self assessment test 9.3

Repeat the appropriate steps to find the collector current when the collector–emitter voltage is 12 V.

Note

You should find that, for $V_{CE} = 12$ V, $I_C = 3.3$ mA.

From these two values of I_C, I can get a value for g_o.

$$g_o = \delta I_C/\delta V_{CE} = 3 \times 10^{-4}/10 = 3 \times 10^{-5} \text{ S}$$

I can get an estimate for β by assuming that base current is entirely due to the emitter efficiency being less than 100 per cent.

Assuming negligible recombination, the ratio of electron current to hole current across the emitter–base junction is given by

$$\frac{I_n}{I_p} = \frac{D_n}{D_p} \times \frac{N_D}{N_A} \times \frac{\text{width of emitter neutral region}}{\text{active base width}}$$

The active base width changes as the collector voltage changes, but taking a representative value of 2.5 μm (corresponding to $V_{CE} \approx 10$ V) gives

$$I_n/I_p = 650$$

Since I_n forms the collector current and I_p must be supplied by the base current

$$\beta = 650$$

Taking the current as nominally 3 mA, $r_e \approx 25/3 \ \Omega$, so

$$r_{b'e} = 650 \times 25/3 = 5.4 \text{ k}\Omega$$

Again, taking the current as 3 mA,

$$g_m = 120 \text{ mA V}^{-1}$$

The value of $r_{bb'}$ depends entirely on the construction of the transistor. Its value is typically $\approx 30 \ \Omega$.

I am now left with the values of two capacitors to calculate. Substitution into Equation (5.17),

$$C_t = A \sqrt{\frac{e\epsilon_0\epsilon_r}{2(1/N_A + 1/N_D)(|\psi| - V_D)}}$$

gives, for a collector–emitter voltage of 2 V, a collector junction capacitance of 3 pF, and for a voltage of 12 V, 1.3 pF. This is $C_{b'c}$.

Using the same formula gives, for the emitter transition region capacitance, 32 pF, and to obtain $C_{b'e}$ we must add to this the diffusion capacitance of charge in the base (there is negligible diffusion capacitance in the emitter because there are relatively few injected holes, and even if the neutral region were long they have a short lifetime).

Equation (5.19),

$$C_{dp} = \frac{l_p^2 I_n e}{2D_n kT}$$

can be applied here to give

$$C_d \approx \frac{W^2 I_C}{2D_n} \frac{e}{kT}$$

giving a value, which again varies with V_{CE}, between 110 pF at 2 V and 102 pF at 12 V. The value of $C_{b'e}$, then, is ≈ 140 pF.

Summarizing, for $V_{BE} = 0.6$ V, $I_C \approx 3$ mA,

$r_{bb'} \approx 30 \ \Omega$

$r_{b'e} \approx 5.4 \ k\Omega$

$g_m \approx 120 \ mA \ V^{-1}$

$g_o \approx 3 \times 10^{-5} \ S$ (output resistance, $r_o = 1/g_o$, $\approx 33 \ k\Omega$)

$C_{b'e} \approx 140$ pF

$C_{b'c} \approx 2$ pF

If V_{CE}, and hence I_C, is increased, then $r_{b'e}$ is reduced, while g_m, g_o and $C_{b'e}$ are increased.

Self assessment test 9.4

From the value 3×10^{-5} S for g_o what is the value of the Early voltage for the transistor?

Answer

Using

$$g_0 = \frac{I_C}{V_{CE} + VA}$$

gives $VA = 98$ V.

9.6 THE BIPOLAR JUNCTION TRANSISTOR AS A SWITCH

To use a BJT as a switch it is normally connected with a resistor between the collector and the supply voltage as shown in Fig. 9.4. With the base–emitter voltage set to zero the transistor passes virtually no current and the output voltage is $+V_{cc}$ – or somewhat below this if the next stage draws current.

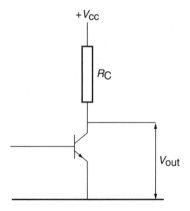

Figure 9.4   A bipolar junction transistor switch circuit.

To switch V_{out} to near zero requires a voltage on the base large enough to pass enough collector current that most of V_{cc} is dropped across R_C. This involves establishing the diffusion gradient in the base, and if this is to happen quickly, charge must be pushed into the base as fast as possible. This can be done by connecting the base to V_{cc} through a resistor which is just large enough to limit the initial base current to a value which will not destroy the transistor. Alternatively the base can be driven by the collector current of another transistor, taking advantage of the fact that a transistor output acts approximately as a constant current generator. This second is the method used in the family of digital gates known as 'transistor–transistor logic' or TTL.

The resistor R_C both lowers the value of V_{CE} and limits the collector current.

As the base current flows, at first it supplies the charge to establish the electron density gradient and to alter the transition region widths, but then, as it goes on flowing, charge builds up in the base until recombination in the base balances the emitter, base and collector currents. In its final condition the charge distribution in the base is as shown in Fig. 9.5. The transistor is said to be in 'deep saturation' and the collector emitter voltage is very small, ≈ 0.1 V. Notice that, using constant-current drive to the base, the value of V_{BE}, and hence the minority carrier density at the emitter junction depletion layer edge, is not directly controlled.

In Fig. 9.5, Q_B may be regarded as the initial switching charge. By the time the transistor has this charge in the base, the base collector junction has zero bias, indicating that (depending on the value of V_{cc}/R_C) $V_{out} \approx 0.6$ V. As the current continues to flow, the extra charge, Q_{BS}, accumulates and V_{out} falls further. It is advantageous that V_{out} should be as small as possible, so from this point of view, saturation is a good thing, but when the transistor requires to be switched off, the extra charge, Q_{BS} (as well as Q_B), has to be removed. Hence, for BJT switches, the time taken to switch from hard-on to off is generally longer than the time to switch from off to effectively on.

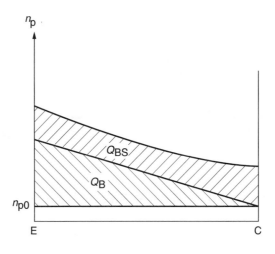

Figure 9.5

A way to reduce the longer switch-off time is to limit the saturation; this is sometimes done by connecting a Schottky diode between the base and collector contacts of the transistor as shown in Fig. 9.6. When the base–collector junction becomes forward-biased 0.3 V, the Schottky diode starts to conduct and diverts further current in the base lead directly to the collector lead so that no further charge accumulates in the base. V_{out} settles down to about $0.6 - 0.3 = 0.3$ V without the transistor being heavily saturated, which is a good compromise. In practice this can be achieved very simply, as illustrated in Fig. 9.7. A layer of aluminium is formed, straddling the collector and base edges of the transistor at

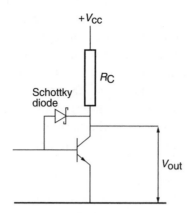

Figure 9.6   A bipolar junction transistor switch circuit with a Schottky diode between base and collector.

the surface. The aluminium forms a rectifying junction with the n-type silicon of the collector, but an ohmic contact with the p-type silicon of the base. Such an arrangement is known as a 'Schottky diode clamp'.

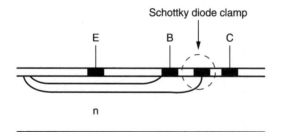

Figure 9.7   A silicon planar npn transistor with a Schottky diode clamp.

9.7   NPN AND PNP TRANSISTORS IN INTEGRATED CIRCUITS

If you want to form a number of npn transistors on the same chip and you do not want them to have all their collectors connected together, some means has to be found to isolate them. This can be done by forming the transistors in an n-type epitaxial layer grown on a p-type substrate. p-type doping is diffused or implanted through the epitaxial layer and the substrate is connected to the negative supply line so that all the substrate–collector junctions are reverse-biased. The arrangement is shown in Fig. 9.8. The bulk resistances of the collectors are made small by depositing 'buried layers' of heavily doped n-type silicon before growing the epitaxial layer, as shown. (The vertical distance is exaggerated in the diagram, so that the distance which the collector current has to travel in high resistance material is short.)

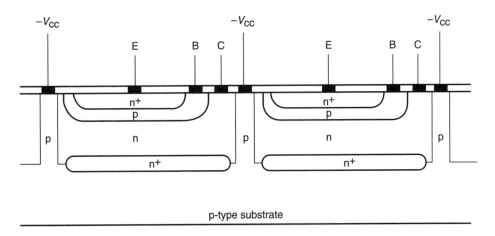

Figure 9.8   Isolation of npn transistors in an integrated circuit.

If pnp transistors are required on the same chip, they can be formed as shown in Fig. 9.9. This structure is known as a 'lateral' transistor. It is difficult to get a narrow enough base width, and the geometry is far from ideal, so lateral transistors have low values of β and generally poor performance compared to the npn transistors. A great deal of ingenuity goes into designing integrated circuits which use this construction, in order to compensate for the vastly different performance of the npn and pnp transistors. If one reversed all the dopings so that the 'good' transistors were pnp, then the npn transistors would be poor, and because of the lower mobility of holes in silicon, the pnp transistors would not perform as well as npn transistors with the opposite dopings.

It is possible to make pnp transistors using the substrate as the collector and the epitaxial layer as the base. They have to be used in circuits, such as the emitter follower, in which the collectors are connected directly to V_{cc}^{-}. The doping levels are not entirely appropriate and it is difficult to achieve a narrow base, so the performance of these 'substrate transistors' is only a little better than that of the lateral pnp transistors.

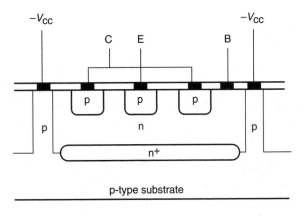

Figure 9.9   A lateral pnp transistor.

9.8 SUMMARY

Base current flows in a BJT with steady bias voltages because:

(a) the emitter efficiency is not 100 per cent;
(b) some recombination occurs in the base.

The value of the current gain, β, $= I_C/I_B$, varies with collector current over a range of about 2 : 1 from the lowest to the highest usable collector currents.
    An effective small-signal equivalent circuit for the BJT is shown in Fig. 9.2:

$g_m = 40\, I_C$ (in mA) mA V^{-1}

$g_o = I_C(V_C + VA)$ and $VA \approx 100$ V or more

$r_{b'e} = \beta r_e \approx 25\beta/I_C$(in mA)   and $\beta \approx 200$ or more

$r_{bb'} = 30\,\Omega$ (typical value)

$C_{b'c} \approx 2$ pF (typical value)

$C_{b'e} \approx 35(1 + I_C$(in mA))pF   ($C_{te} + C_d$, typical values)

    The BJT, when used as a switch, is best current-driven so that the charge necessary to change the state of the transistor enters the base quickly. When the base is current-driven, extra charge accumulates in the base during the 'on' state, so that the time to switch to the 'off' state from the 'on' is extended. Switching from 'on' to 'off' can be speeded up by connecting a Schottky diode clamp, i.e. a Schottky diode between the base and collector, to prevent the transistor from going into deep saturation.
    Integrated circuits involving BJTs often use an n-type epitaxial layer on a p-type substrate so that efficient npn transistors can be made, and these can be isolated from one another. pnp transistors included in such a construction are far less effective than the npn transistors, and the designs of the circuits used have to take account of this.

9.9 PROBLEMS

1   Consider a transistor, identical in dimensions and doping densities to that shown in Fig. 9.3, but which is a pnp device. For the same values of applied voltages (but reversed in polarity of course), deduce which of the parameters calculated in Section 9.5 (p. 124) would be different and calculate their values for the pnp device.

2   A silicon planar pnp transistor has uniform doping in its regions of 5×10^{24} acceptors/m³ in the emitter, 10^{22} donors/m³ in the base and 5×10^{21} acceptors/m³ in the collector. The emitter junction is 6 μm below the surface of the silicon and the collector junction is 4 μm below that. At an operating

point where $V_{BE} = -0.60$ V and $V_{CE} = -5$ V, the values of I_C and I_B are measured as -0.99 mA and $-4.5\,\mu$A, respectively. Calculate what percentage of the holes that are injected into the base from the emitter reach the collector.

10 Light absorption and emission in semiconductors

I am using the term 'light' here to mean electromagnetic radiation of any wavelength between about $0.4\,\mu$m, which is violet light, and about $4\,\mu$m, in the infrared. Only the range between about 0.4 and $0.8\,\mu$m is visible.

It is common practice when discussing light to describe its wavelength rather than its frequency, as I have done above, although frequency is the invariant property in any linear medium. The wavelength meant is the so-called 'free-space' wavelength, λ_0, which is the wavelength of a plane wave of the given frequency in a vacuum; the relationship of this to the frequency is

$$\lambda_0 = c/f$$

where $c = 3 \times 10^8$ m s^{-1}. The reason that free space wavelengths are quoted is that, for light, wavelength can be measured directly and frequency must then be deduced from this.

Self assessment test 10.1

What is the range of frequencies which correspond to the range of wavelengths that I quote in the first paragraph?

Answers

$0.4\,\mu$m corresponds to 7.5×10^{14} Hz; $4\,\mu$m corresponds to 7.5×10^{13} Hz.

10.1 QUANTUM NATURE OF LIGHT

It has been understood for over a hundred years that light is a form of electro-magnetic radiation – a wave of mutually generating electric and magnetic fields propagating through space. The launching of, and the extraction of energy from, electromagnetic waves at frequencies up to at least 10^{10} Hz is organized using antenna theory and does not involve any consideration of quantum effects. However, at frequencies in the 10^{14} Hz range and above, these effects are central in emission and absorption theory.

When electromagnetic radiation of a particular frequency is emitted, the energy of the wave increases in steps or 'quanta' of a size given by the equation

$$\delta E = hf$$

where f is the frequency and h is a physical constant known as Planck's constant whose value is given in Table 1.1. Experiment suggests that the extra energy originates from a small volume in space, often a single atom. When energy is absorbed from an electromagnetic wave it is similarly absorbed in discrete quanta, and again an individual quantum of energy seems to disappear from the wave in a small region. This experimental evidence leads us to model light as being composed of particles of energy, each of magnitude hf, called 'photons'. The photon model does not invalidate the model of light as a wave of electric and magnetic fields – a model which has such wide application at lower frequencies – and somehow one has to accommodate both in one's mental scheme. I should like to emphasize that the only direct evidence for the existence of photons is in absorption and emission, although they form a fundamental part of modern theories of particle physics.

Self assessment test 10.2 _____

Calculate the energy (in joules) of a photon of electromagnetic radiation: (a) of frequency 20 MHz (used for short-wave radio); (b) of frequency 3×10^9 Hz (a radar frequency); (c) of frequency 4×10^{14} Hz (visible red light).

Answers

(a) At 20 MHz, photon energy $= 1.2 \times 10^{-23}$ J; (b) at 3×10^9 Hz, photon energy $= 1.7 \times 10^{-21}$ J; (c) at 4×10^{14} Hz, photon energy $= 2.3 \times 10^{-16}$ J.

Radar receivers can detect microsecond pulses with peak power down to about 10^{-12} watts; at this power level there are still about 600 photons per pulse. Hence the 'graininess' at this and lower frequencies is not generally noticed, in the same way that one does not normally notice that water consists of molecules when considering the flow of the liquid. The same sensitivity at optical frequencies would be impossible, because it would involve considerably less than one photon per microsecond pulse.

10.2 ABSORPTION OF LIGHT BY A SEMICONDUCTOR

Light energy is absorbed by a solid when the photons are able to move electrons up to higher energies in the band structure. In a semiconductor, whether intrinsic or extrinsic, there are always some electrons in the conduction band and some holes in the valence band, so that there are always some electrons that can absorb photons, of however small an energy, by moving up within an energy band. However, there is a great increase in absorption for radiation of a frequency such that the photons have energy greater than the energy gap and can thus promote electrons from the valence band to the conduction band.

As light passes through a material, each photon has a certain probability of being absorbed in each unit of distance travelled, hence the number of photons absorbed per unit distance is proportional to the number present. Mathematically, this leads to the result that

$$N_x = N_0 \exp(-\alpha x)$$

where N_0 is the number of photons entering the material and N_x is the number surviving at distance x. Since light intensity is proportional to the number of photons, the formula can also be expressed in terms of light intensity. α is the 'absorption coefficient' and has dimensions m^{-1}.

If $x = 1/\alpha$, that is $\alpha x = 1$, then

$$N_x = N_0/\exp(1) = 0.37\, N_0$$

so, about two-thirds of the light is absorbed in a distance equal to $1/\alpha$.

Graphs of absorption coefficient against wavelength are plotted in Fig. 10.1 for silicon, for gallium arsenide and for certain alloys of gallium indium and arsenic and of gallium indium arsenic and phosphorus. The reason why I am interested in these last two will become clear later. Three of these materials have sharp 'absorption edges', i.e. wavelengths above which there is negligible absorption and the material is effectively transparent. Absorption in silicon comes on rather more gradually as wavelength falls beyond the cut-off value.

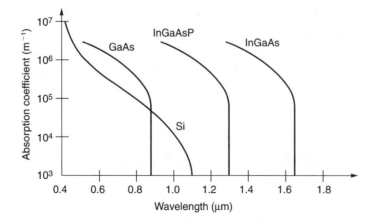

Figure 10.1 Variation of the optical absorption coefficient with wavelength for several semiconductor materials.

Looking at gallium arsenide and the other alloys of gallium you will see that all three have absorption coefficients, for wavelengths which are just below the cut-off value, of something like 5×10^5 m^{-1}. The inverse of this value is about 2×10^{-6} m, indicating that the bulk of the absorption of light entering the material takes place over a distance of a few micrometres.

Transparency and the speed of light in a semiconductor

From what I have said so far you will have gathered that a semiconductor (or insulator for that matter) is deemed to be transparent to light of which the photon energy is insufficient to move electrons from the valence band to the conduction band.

Self assessment test 10.3 _____

From Fig. 10.1, deduce the energy gap in GaAs and compare your answer with the value given in Table 1.2 on page 4.

Answer

From Fig. 10.1 the wavelength of the absorption edge appears to be about 0.875 μm. Photons at that wavelength have energy

$$hc/\lambda_0 = 4.14 \times 10^{-15} \times 3 \times 10^8/0.875 \times 10^{-6} \text{ eV}$$

$$= 1.42 \text{ eV}$$

(Notice that I was able to get the answer directly in electron volts by using the value of h quoted in eVs.) This agrees with the value given in the table.

_____

For frequencies at which the material is transparent there is still some loss of light, by absorption due to electron energies being raised within bands and by scattering of the light from impurities and dislocations.

In a transparent material the electromagnetic wave travels slower than in a vacuum. This 'slowing down' can be explained in terms of an interaction between the photons and the electrons in the valence band, in which the photons are absorbed and re-radiated with a delay. The ratio of the speed in a vacuum (c) to the speed of the light in the material is known as the 'refractive index' of the material. For frequencies at which the material absorbs radiation, those photons which at any stage are not absorbed are slowed down in a similar way.

Electromagnetic theory shows that the refractive index of a semiconductor or insulator is equal to the square root of its relative permittivity. The value of refractive index varies somewhat with frequency, and thus, by implication, so does the value of relative permittivity – values quoted in data sheets are usually low frequency values. For semiconductor materials the refractive index is in the range 3–4 (for comparison, glass has a refractive index of about 1.5).

Since when light enters a material its frequency is unchanged (except in some special circumstances where the medium is non-linear) the wavelength in the material is shorter than the free space value. I can write

$$\lambda_m = \lambda_0/n$$

where n is the refractive index for the frequency concerned.

10.3 EMISSION OF LIGHT BY A SEMICONDUCTOR

A carrier electron and a hole are formed when an electron moves from the valence band to the conduction band; the energy for this can come from heat, but it can also be provided by a photon, as we have seen. Conversely, when an electron and a hole meet and recombine, the energy released may be in the form of heat or it may emerge as a photon of light, at a frequency determined by the amount of energy. Generally, both processes occur, but in some materials, silicon being a good example, hardly any of the energy is liberated as light. In other materials, half, or sometimes more, of the recombinations result in the release of a photon. This difference in behaviour is primarily determined by whether the material has an indirect or a direct energy gap.

Direct and indirect energy gaps

The energy difference between an electron in the conduction band and one in the valence band is primarily potential energy: an electron in the conduction band is, on average, farther from the positively charged core atoms than is an electron in the valence band. Moving up the conduction band for an electron,

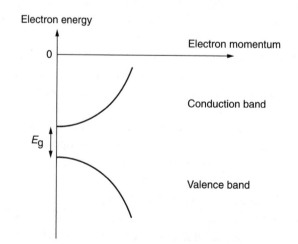

Figure 10.2 Electron momentum versus energy for an ideal semiconductor.

or down the valence band for a hole, implies adding kinetic energy to the potential energy of its general location in the structure. Giving an electron kinetic energy means also giving it momentum, and Fig. 10.2 is the graph that I would expect to relate electron momentum to electron energy for a material with precisely these properties. Practical materials are more complicated than this, and for some, the lowest energy in the conduction band does not coincide with zero momentum; in other words, some of the energy of the very lowest energy conduction band electrons is kinetic.

Figure 10.3 shows graphs of electron energy versus momentum for silicon and for gallium arsenide. The detail depends on the direction in which the electrons are moving in the crystal structure, and in fact the very lowest energy conduction electrons in silicon are moving in the direction of one specific crystallographic axis. However, such detail aside, in gallium arsenide, both a hole

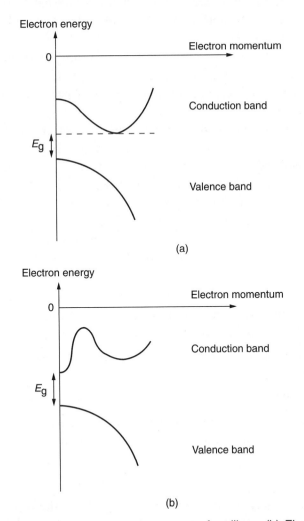

Figure 10.3 (a) Electron momentum versus energy for silicon. (b) Electron momentum versus energy for gallium arsenide.

at the top of the valence band and electron at the bottom of the conduction band have zero momentum; such a material is described as a 'direct energy gap' semiconductor. In silicon, again the holes at the top of the valence band have zero momentum, but the lowest energy electrons in the conduction band have a significant momentum, so silicon is described as an 'indirect energy gap' semiconductor.

Radiative and non-radiative recombination

In any interaction between two particles, both energy and momentum must be conserved. When an electron at the bottom of the conduction band in gallium arsenide meets a hole at the top of the valence band and tries to recombine with it so that both particles disappear, energy has to be carried away from the point of recombination, but there is no momentum mismatch. If, on the other hand, the electron and the hole are in silicon, then the electron has momentum and this must be disposed of.

Photons of light have energy but little momentum (light does have some momentum, as can be shown classically as well as by quantum mechanics), so that the recombination in gallium arsenide can easily produce a photon of light to remove the energy.

Momentum can be taken up by the vibrations of the lattice associated with heat, and energy can also be removed in this way. In silicon, a photon will only be emitted if its emission is coordinated with the coupling of vibrational energy to the lattice. This is an unlikely event, so very few photons are emitted. In fact, the removal of both energy and momentum in the correct proportions by coupling directly to lattice vibrations is not very likely either, and so recombination in silicon depends quite a lot on impurities and dislocations which 'trap' the electrons and holes and cause them to recombine in stages.

[*Note*   In a quantum mechanical treatment, lattice vibrations can be modelled as particles called 'phonons'. A phonon carries away from a recombination a little energy and a lot of momentum. I shall not pursue this analysis.]

Self assessment test 10.4 _____

Following from the above analysis, can you offer an explanation as to why the absorption edge in silicon, shown in Fig. 10.1, is not sharp?

Explanation

Electrons cannot easily be lifted by light energy from the top of the valence band and enter energy levels at the bottom of the conduction band because of the change of momentum required. As we move to shorter wavelengths, i.e. higher frequencies, the photons have enough energy to lift electrons to energy levels which have little or no momentum associated with them, so the absorption increases more gradually as one goes down through the wavelength associated with the bandgap energy than is the case for the other materials shown.

The likelihoods of a radiative and a non-radiative recombination, respectively, occurring in a given material are incorporated into two carrier lifetimes: t_{rr}, the radiative lifetime, and t_{nr}, the non-radiative lifetime. These two lifetimes are defined as follows. If an excess of electrons and holes is introduced into the material and if the non-radiative process did not exist, then the average time that a carrier exists before it recombines and emits a photon is the radiative carrier lifetime. The non-radiative lifetime is the average lifetime of an excess carrier if the radiative process did not exist. There are means by which the two lifetimes in a given material can be measured.

I can now define an *internal quantum efficiency*, η_{int}, as the proportion of electron-hole recombinations which result in photon emission. The number of radiative recombinations per second will be $1/t_{rr}$, while the total number of recombinations per second must be $1/t_{rr} + 1/t_{nr}$ so

$$\eta_{int} = \frac{1/t_{rr}}{1/t_{rr} + 1/t_{nr}}$$

$$= \frac{1}{1 + t_{rr}/t_{nr}}$$

Thus, for efficient light emission, the radiative lifetime must be much shorter than the non-radiative lifetime.

10.4 APPROPRIATE SEMICONDUCTOR MATERIALS FOR OPTICAL USE

Silicon and germanium are both indirect energy gap semiconductors; the direct energy gap materials are all compound semiconductors. Most compound semiconductors are made from a combination of elements from group 3a and group 5a of the periodic table of elements.

[*Note* It is possible to make semiconductor material from combinations of group 2b and group 6a elements, but I shall not consider these.]

Table 10.1   Extract from the periodic table showing the atomic number and group of relevant elements

Groups	2b	3a	4a	5a	6a
		5 B	6 C	7 N	
		13 Al	14 Si	15 P	16 S
	30 Zn	31 Ga	32 Ge	33 As	34 Se
	48 Cd	49 In	51 Sb	51 Sb	52 Te

Table 10.1 shows part of the periodic table which contains those elements that find a use in the construction of semiconductor devices. The designation 'a' or 'b' is relevant to chemists and indicates details of the electron configurations in the atoms. From our point of view, the important idea is that the group 2 atoms contribute two electrons to valence bonds in a crystal, group 3 atoms contribute three electrons, group 4 atoms four, group 5 atoms five and group 6 atoms six.

Both the single-element semiconductors are from group 4 and are doped to make them p-type or n-type with group 5 and group 3 atoms, respectively. When just group 3 and group 5 materials are mixed, crystals form with an exact balance of three-valent to five-valent material so that there is an average of four valence electrons per atom, and the structure is similar to the silicon crystals described in Chapter 3. To dope such a material p-type, the most effective way is to add a small amount of group 2 material: the group 2 atoms replace some of the group 3 atoms releasing holes. To achieve n-type doping, group 6 atoms are used and take the place of some of the group 5 atoms, releasing electrons.

10.5   SUMMARY

Light is taken to mean visible and near-infrared radiation. It is usually specified by its free-space wavelength rather than its frequency because its wavelength can be directly measured.

Electromagnetic radiation is emitted and absorbed in quanta. This means that the radiation can be described as consisting of photons of energy, each photon having an energy content which depends on its frequency, given by

$$E_{ph} = hf$$

One way of reconciling this model to the description of electric and magnetic fields as waves is to think of a photon as a small localized wave packet.

A photon at light frequencies has a large energy content, in atomic terms, but a very small momentum.

A semiconductor is regarded as transparent to light whose frequency is such that the photon energy is less than its energy gap – although there will still be some attenuation of light passing through. For light with a photon energy greater than the energy gap in the material, the light is absorbed over the distance of a few micrometres in the material. The velocity of light in a semiconductor is reduced compared to the free-space velocity, and hence the material wavelength is less than the free-space wavelength. The ratio of the free-space velocity to the velocity in the material, and hence also the ratio of the free-space wavelength to the wavelength in the material, is the refractive index n.

The single-element semiconductors, silicon and germanium, have indirect energy gaps in which the momentum of holes at the top of the valence band is negligible, but the electrons at the bottom of the conduction band have significant momentum. This makes it unlikely that an electron and hole combining will produce a photon. Many of the semiconductors formed as alloys of elements from groups 3 and 5 of the periodic table have direct energy gaps with negligible momentum of both holes near the top of the valence band and electrons near the bottom of the conduction band; in such materials radiative recombination is much more likely. The internal quantum efficiency of a semiconductor, that is the proportion of recombinations that result in the emission of a photon, is given by the formula

$$\eta_{int} = \frac{1}{1 + t_{rr}/t_{nr}}$$

where t_{rr} is the radiative lifetime and t_{nr} the non-radiative lifetime of carriers.

The doping of 3–5 alloy semiconductors is achieved by using group 2 atoms as acceptors and group 6 atoms as donors.

10.6 PROBLEMS

1 Deduce the energy gaps in the two alloys of indium gallium arsenide phosphide and indium gallium arsenide whose absorption characteristics are shown in Fig. 10.1.

2 If, for light of a particular frequency, a semiconductor has an absorption coefficient of 5×10^5 m^{-1}, calculate the distance that light must pass through the material for 95 per cent to be absorbed.

3 What is the frequency of light which, in a semiconductor of refractive index 3.5, has a material wavelength of 0.4 μm?

4 What is the internal quantum efficiency for light production of a semiconductor material in which the radiative lifetime is 10^{-7} s and the non-radiative lifetime is 2×10^{-7} s?

11 Photocells and photodiodes

This chapter is concerned with means for turning light signals into electrical signals. Perhaps the most obvious way of causing light to change an electrical parameter is to allow the light to penetrate a suitable piece of semiconductor in which it will create electron–hole pairs and thus change the resistance; this can be used as a means of detecting infrared radiation (for example) However, here I am only going to consider devices which incorporate a pn junction.

11.1 CHARACTERISTICS OF A PN JUNCTION THAT ABSORBS LIGHT

If light of wavelength smaller than the cut-off wavelength enters a semiconductor, it is absorbed over a distance of a few micrometres, and electron–hole pairs are created. If the dimensions are appropriate and the light enters in such a way that it is absorbed either in the transition region or in the neutral regions within a diffusion length of the transition region, then the electron–hole pairs created result in an extra current, over and above the normal current, flowing in the diode. In practice, devices are designed so that as far as possible the carrier generation occurs in the transition region, so I shall assume that that is the case.

Imagine an electron and a hole created together somewhere in the transition region of a pn junction. They find themselves in an electric field which propels the hole into the p region and the electron into the n region. Figure 11.1 illustrates simply that the current generated by an electron and a hole which are created and separated in the way I have described is the same as the current due to a single charge passing straight through the junction.

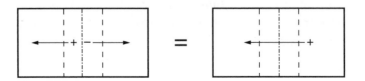

Figure 11.1

Not every photon absorbed results in a permanent electron–hole pair that can separate as described – some pairs recombine, turning the photon energy into heat – so I need to define a quantum yield, η, which is the fraction of absorbed photons which result in units of electronic charge crossing the junction.

[**Note**   This parameter, η, is also sometimes called quantum efficiency, but since I have already used that term in the context of emission (see Chapter 10), I shall stick with quantum yield.]

Since the normal junction current is defined as a forward current when it flows from the p side to the n side, the photocurrent (optically generated current) I_p, in the junction subtracts from the normal junction current, and I can write, for the total current,

$$I_D = I_s[\exp(eV_D/kT) - 1] - I_p$$

Assuming that the quantum yield remains the same, equal increments of light intensity (at a fixed wavelength) will give equal increments of I_p; the junction characteristics will be as shown in Fig. 11.2.

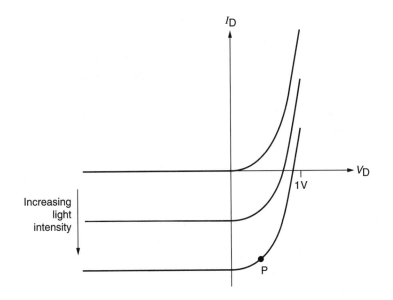

Figure 11.2   Photo-absorption characteristics of a pn junction.

11.2   PHOTOCELLS

If a diode containing a pn junction with an appropriate energy gap, into which light can penetrate, is simply connected across a load, then the diode will act as a source of emf and drive a current through the load. The source of energy is of course the light. Figure 11.3 shows a circuit diagram and indicates the directions of the voltage and current. The diode is operating at a point on the characteristics, marked P in Fig. 11.2, at which the current and voltage are V and I as shown in Fig. 11.3. The position of the point P depends on the load resistance, which must be selected so that the operating point sits on the 'knee' of the characteristic curve. Typically the voltage associated with point P will be around 0.4 V. Such a diode is a photocell and is said to be operating in the *photovoltaic* mode.

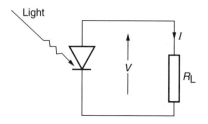

Figure 11.3

If a small photocell is required to respond to daylight, then single-crystal silicon will do as the semiconductor material. The absorption characteristic indicates that all the visible range of daylight is absorbed effectively.

Photocells to convert sunlight into significant amounts of electrical energy need to have a large area so as to capture as large an amount of light as possible. To make a large area device cheaply is difficult if single-crystal material is to be used, so one solution is to use an amorphous silicon–hydrogen alloy. Amorphous silicon is like glass – it does not have a regular crystal structure, but is nonetheless homogeneous – and it is relatively easily formed in thin layers by depositing it, usually on the metal which is to form one contact, from a vapour. Amorphous silicon has a rather ill-defined band gap, but it can be doped n-type or p-type with the usual dopants, although the doping densities need to be rather large. Because of the irregularities in the internal structure there are many dangling bonds in pure amorphous silicon, and these form traps and reduce the mobility of carriers dramatically. However, it has been discovered that adding hydrogen when the material is formed terminates these dangling bonds and minimizes their effect.

To make sure that there is a wide transition region in which the electron–hole pairs are to be formed, the structure is doped as pin, that is, an intrinsic layer of material is sandwiched between the p-doped side and the n-doped side as shown in Fig. 11.4. In such a junction, when the initial diffusion of carriers sets up the transition region, it extends right across the intrinsic material, and since the electric field in a transition region is due to the uncovered dopant ions on either side, in this case there is electric field through the whole intrinsic region. Since the doping in the p and n materials is heavy, the transition region does not extend very far into the p and n regions, so in effect the transition region coincides with the intrinsic part of the structure.

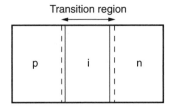

Figure 11.4 A pin junction.

Amorphous silicon solar cells can have energy conversion efficiencies of 10–15 per cent. Because they are relatively cheap to make, amorphous silicon solar cells are often used even for low energy applications such as powering a watch or a calculator.

11.3 PHOTODIODES

A very important application of the conversion of light signals into electric signals is at the receiving end of a telecommunications optical link. Pulses of light arrive at a very rapid rate (up to 10^{10} pulses per second) on an optical fibre and must be converted into electrical pulses. This requires a device that is sensitive, linear and responds extremely rapidly; experience shows that a diode operating in the photovoltaic mode is not appropriate. A photodiode that is reverse-biased will act as a controlled current generator, in the same way that a transistor does, but with the controlling input being the light. Looking at the reverse voltage part of the characteristics shown in Fig. 11.2 may help you to understand this. A diode used in this way is said to be operating in the *photoconductive* mode, and optical receiver photodiodes are designed to be used like this, with a resistor in series across which a signal voltage will be developed.

A useful parameter to define for a photodiode is the responsivity, R. This is the photocurrent per unit incident light power

$$R = \frac{I_p}{P}$$

The number of photons per second in light of power P is

$$\frac{P}{hf} = \frac{P\lambda_0}{hc}$$

and thus the number of electron–hole pairs created per second is

$$\frac{\eta P\lambda_0}{hc}$$

Each of these carrier pairs contributes a unit of electronic charge to the photocurrent, so

$$I_\mathrm{p} = \frac{e\eta P\lambda_0}{hc}$$

and thus finally

$$R = \frac{\eta e\lambda_0}{hc}$$

This formula does not apply, of course, if the photon energy is less than the energy gap in the semiconductor material; for a given diode there is a maximum value of λ_0.

In all materials, as one moves from the cut-off wavelength to shorter wavelengths the absorption coefficient increases. If, as is normally the case in communications applications, the light is at one wavelength (or occupies a small band of wavelengths), then since it is important to get the light into the diode to where you want it, it is generally necessary to choose a diode made of a material with a cut-off wavelength not too much longer than the wavelength of the light.

Two narrow (in percentage terms) wavelength bands are used in optical communications, centred on $\lambda_0 = 1.3\ \mu\mathrm{m}$ and $\lambda_0 = 1.55\ \mu\mathrm{m}$. These wavelengths are dictated by the transmission properties of silica fibre; at the shorter wavelength the fibre has least distorting effect on pulses, and at the longer wavelength the fibre has minimum attenuation. To choose a suitable material for a photodiode to receive radiation in both these bands I can first examine the energy gaps of some of the more easily manufactured semiconductor materials that are available (see Table 11.1).

Table 11.1   Energy gaps and cut-off wavelengths of some semiconductors

Semiconductor material	Direct or indirect gap (D or I)	Energy gap (eV)	Cut-off wavelength (μm)
Indium antimonide (InSb)	D	0.17	7.29
Indium arsenide (InAs)	D	0.36	3.44
Germanium (Ge)	I	0.67	1.85
Gallium antimonide (GaSb)	D	0.72	1.72
Silicon (Si)	I	1.12	1.11
Indium phosphide (InP)	D	1.35	0.92
Gallium arsenide (GaAs)	D	1.42	0.87
Gallium phosphide (GaP)	I	2.26	0.51
Silicon carbide (SiC)	I	3.00	0.41

The two materials that would do are germanium and gallium antimonide; both of these, however, have energy gaps smaller than necessary. Why should this matter?

Besides the photocurrent, the reverse-biased diode has a 'dark current', so called because it flows in the absence of any light. This dark current is none other than the reverse saturation current of the diode, I_s, which is larger the smaller the energy gap.

Self assessment test 11.1_____

Why is I_s larger the smaller the energy gap, all other things being equal?

Explanation

I_s depends on minority carrier densities (see Equation (5.11) of Chapter 5). These depend on n_i which in turn depends on the energy gap.

The dark current contributes to noise in the diode without contributing to the signal, so it needs to be as small as possible, which indicates that the cut-off wavelength of the diode material should be longer than the signal wavelength by as small an amount as is practical.

Indium gallium arsenide

Indium gallium and arsenic amalgamate easily together and can form crystals with any proportions of gallium to indium, the sum of the two (both trivalent) being equal to the amount of arsenic (which is pentavalent). The bandgap of the material varies smoothly with composition from 0.36 eV for pure indium arsenide to 1.42 for pure gallium arsenide. The bandgap associated with a cut-off wavelength of 1.3 μm is 0.95 eV, and that for 1.55 μm is 0.8 eV, so it is possible to tailor an appropriate indium gallium arsenide alloy for each wavelength. There is, however, a snag. Indium gallium arsenide with accurate predictable proportions of indium to gallium, and with the required doping, cannot be grown in bulk; it has to be laid down by liquid epitaxy (or some similar method). On what can it be deposited? The answer is some other semiconductor that can be grown in bulk. Here is another problem. To grow an epitaxial layer of one material on another successfully, the crystal lattices must 'line up'. The length of the side of a unit cell of a crystal is called the 'lattice parameter' for that crystal. If the lattice parameters of two materials with similar crystal structures are equal within less than 0.1 per cent or so, then an interface will form between them without too many strains or dislocations to disrupt the flow of carriers and reduce lifetimes.

Different compositions of indium gallium arsenide have different lattice parameters as well as different energy gaps. The most appropriate substrate material proves to be indium phosphide, and this is lattice-matched to indium gallium arsenide of composition 53 per cent indium, 47 per cent gallium, usually designated $In_{0.53}Ga_{0.47}As$. This composition has an energy gap of 0.75 eV, corresponding to a cut-off wavelength of 1.65 μm, which is taken as good enough for 1.3 μm and 1.55 μm receivers.

The indium gallium arsenide/indium phosphide heterojunction

For the first time we meet here a new situation: two different semiconductors, with different band gaps, are lattice-matched and formed into a continuous crystal structure. Suppose the indium phosphide is doped N-type and the indium gallium arsenide p-type (it is normal practice to designate the doping of the wider-gap material with a capital letter); how will the junction behave?

It is instructive to draw energy band diagrams, first for the materials separately and then together (see Fig. 11.5). For the materials joined, the Fermi levels come into line by the initial exchange of charge, and the band edges bend, with discontinuities in the band edges at the junction. There is a

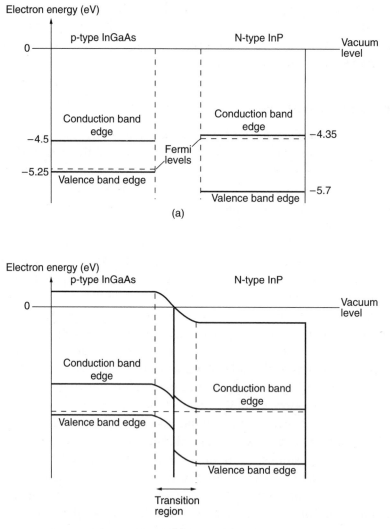

Figure 11.5 Energy band diagrams for an indium gallium arsenide indium phosphide heterojunction: (a) materials separate; (b) materials joined.

transition region, as in a single-material junction (a homojunction). When the junction is forward-biased, many electrons flow from the N to the p region – the situation in the conduction band is effectively the same as it would be for a homojunction of indium gallium arsenide – but very few holes flow from p to N because the valence band edge in the N material is still far below that in the p material and there are very few holes so low in the p-material valence band. When the junction is reverse-biased, again the current flow is almost entirely made up of electrons because the N material, having a wide gap, has very few minority carriers. The general conclusion, which applies to all heterojunctions, is that junction current is carried almost entirely by the type of carrier which is the majority carrier in the widegap material.

An indium gallium arsenide/indium phosphide piN photodiode

Indium phosphide, with a cut-off wavelength of 0.92 μm, is transparent to radiation at 1.3 μm and at 1.55 μm, so if a layer of p-type indium gallium arsenide were formed on a substrate of indium phosphide to make a photodiode, all the carrier pair generation would have to take place in the p-type material. To give the device a fast response it would be necessary to try to arrange that most of the photon absorption takes place in the p side of the transition region – any electrons generated in the neutral p region would have to diffuse to the transition region, which is a slow process. There is also a problem with the junction capacitance, which for bandwidth reasons needs to be kept both small and predictable. All these considerations lead to a design in which the indium gallium arsenide has a junction in it between heavily doped p$^+$ material and lightly doped n$^-$ material. The indium phosphide substrate is N-type, and again heavily doped. A suitable structure is shown in Fig. 11.6.

If sufficient reverse bias is applied, the n$^-$ material is completely depleted

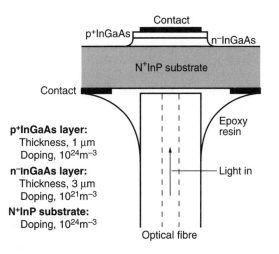

Figure 11.6 Structure of a piN photodiode.

and there is very little of the transition region in the p$^+$ or the N$^+$ material. The light enters through the substrate, which is transparent to it, and the n$^-$ region is wide enough that effectively all the light is absorbed there. The capacitance is small and predictable, being determined effectively by the material properties and width of the n$^-$ layer, and the current falls rapidly when the light intensity falls because the carriers are removed from the n$^-$ region by the transition region electric field. Such a device is usually called piN, but the middle of the sandwich cannot actually be made intrinsic because its carrier levels would be too unpredictable.

Self assessment test 11.2

Given the dimensions and doping levels shown in Fig. 11.6, calculate the minimum reverse voltage required across the diode in operation. You will need to know n_i for indium gallium arsenide at room temperature, which is $\approx 10^{18}$ m^{-3}, and ϵ_r for the material, which is 14.

Solution

The voltage has to be large enough to fully deplete the 'i' (n$^-$) layer. Effectively all the transition region is in the i region, so taking its width as the width of the transition layer when it is just depleted and using Equation (5.15),

$$w_t = \sqrt{\left[\frac{2\epsilon_0\epsilon_r}{e}\left(\frac{1}{N_A} + \frac{1}{N_D}\right)\left(|\psi| - V_D\right)\right]}$$

$$-V_D = \frac{w_t^2 e}{2\epsilon_0\epsilon_r(1/N_A + 1/N_D)} - |\psi|$$

$$= \frac{(3 \times 10^{-6})^2 \times 1.6 \times 10^{-19}}{2 \times 8.85 \times 10^{-12} \times 14 \times (10^{-24} + 10^{-21})} - |\psi|$$

$$= 5.8 - |\psi|\ \text{V}$$

I don't really need to calculate $|\psi|$; I know that it will be a fraction of a volt, but just to check:

$$|\psi| = |(kT/e)\ \ln(n_i^2/N_A N_D)| = |0.025\ \ln(10^{36}/10^{45})|$$

$$= 0.52\ \text{V}$$

The voltage across the diode must be at least 5.3 V.

To give the photodiode a fast response the carriers need to move out of the transition region as quickly as possible. This suggests as large a transition region electric field as possible; however, as mentioned in Chapter 5, carrier velocities saturate at a field strength of $\approx 10^6$ V m^{-1}, so there is no extra speed of response gained by exceeding this value.

Self assessment test 11.3 _____

Does the electric field in the i region with an applied voltage of 5.3 V reach the saturation value?

Answer

Most of the voltage is across the depleted i region, so the average field is given by $E_{av} \approx 5.3/3 \times 10^{-6} = 1.77 \times 10^6$ V m^{-1}.

As we saw in Chapter 6, avalanche breakdown is likely to occur with transition region electric fields of the order of 5×10^7 V m^{-1}. To produce an average field of half of this (which means a peak field of the value) requires a voltage of

$$2.5 \times 10^7 \times 3 \times 10^{-6} = 75 \text{ V}$$

So, the diode can safely be operated with a bias of, say, 10 V.

Sometimes piN diodes are designed to be operated near avalanche breakdown so that the initial photocurrent is multiplied by avalanching inside the device. Such 'avalanche photodiodes', however, suffer from greater noise than those operated below a voltage at which avalanching is significant.

11.4 SUMMARY

When light is allowed to enter the transition region of a pn junction in material with a cut-off wavelength longer than the wavelength of the light, a fraction, η, of the photons create electron–hole pairs; η is called the quantum yield. The electron and the hole are driven by the transition region electric field into the neutral n region and the neutral p region, respectively, and the effect is equivalent to one unit of electronic charge crossing the junction in the reverse-current direction. This photocurrent adds negatively to the normal diode current producing characteristics as illustrated in Fig. 11.2.

Semiconductor diodes are used to respond to light in two ways. In the photovoltaic mode, an unbiased diode, operating over the forward-current region of the junction characteristics, acts as a source of emf, turning light energy into electrical energy; this is the function of a solar cell. In the photoconductive mode, a reverse-biased diode acts as a constant-current generator, releasing current proportional to the incident light intensity; a communications photodiode acts in this way.

Solar cells are often made from a Si:H alloy and incorporate a pin structure so as to make the transition region wide enough for efficient light absorption.

Photodiodes for long-distance communications are designed for light wavelengths of around either 1.3 μm or 1.55 μm. The most suitable material is indium gallium arsenide, but this cannot be made in bulk, so it is grown on an indium phosphide substrate. To grow the indium gallium arsenide successfully it has to be lattice-matched to the indium phosphide and this constrains its composition to one with a cut-off wavelength of 1.65 μm, which is found to be satisfactory for both signal wavelengths. The structure used is a sandwich of n$^-$ InGaAs between p$^+$ InGaAs and N$^+$ InP, forming what is usually described as a piN structure (despite the fact that the i section is really n$^-$ material).

In use, a piN diode requires a bias voltage of 10 V or so, although diodes of a similar structure are sometimes designed for a much higher bias voltage, causing avalanche gain inside the diode but also more noise.

The photocurrent per unit light power is called the responsivity of the photodiode and is given by

$$R = \frac{\eta P \lambda_0}{hc}$$

11.5 PROBLEMS

1 A solar cell of area 1 cm^2 is exposed to intense sunlight with a power flux of 1 kW m^{-2}. The value of I_s for the cell junction is 1 nA and the quantum yield in the transition region is 50 per cent. The cell is to be connected across a suitable load such that the voltage drop across the load is 0.38 V. Taking the average wavelength of the sunlight to be 0.7 μm, estimate:
   (a) the current generated;
   (b) the required load resistance;
   (c) the conversion efficiency of the cell.

2 A piN photodiode, as described in the text, is connected in series with a 1 MΩ resistor and a voltage supply of 10 V which reverse biases the diode. The quantum yield in the 'intrinsic' region of the diode is 67 per cent. Light pulses of wavelength 1.3 μm and peak power 1 μW are received by the diode. Deduce:
   (a) the pulse voltage appearing across the resistor;
   (b) the peak power dissipated in the diode.

12 LEDs and lasers

LEDs are light-emitting diodes and 'laser' is an acronym for 'light amplification by stimulated emission of radiation'. Although laser is an acronym it is usually written in lower case, and despite the acronym most lasers are oscillators and hence sources of radiation. Both types of device are based on a forward-biased pn junction made from semiconductor material with a short radiative carrier lifetime and thus a high internal quantum efficiency. Usually this means using a direct energy gap material, but not always.

To get a significant light output there needs to be a large excess of holes and electrons ready to recombine. An obvious way to achieve this in a semiconductor is to forward bias a pn junction. In a rectifying diode we want the excess carriers not to recombine, so that the minority carriers can diffuse away to the contacts; here we want them to recombine, radiatively, as soon as possible. The process of getting light this way is called *injection luminescence.*

Since most of the electrons are at the bottom of the conduction band and most of the holes are at the top of the valence band, the most likely wavelength to be emitted is that for photons of the bandgap energy; however, some of the carriers are away from the band edges, and also some of the recombinations can result in the production of heat energy as well as light, so there is an energy spread and thus a wavelength spread in the output.

To produce a given central emission wavelength, and hence a given colour if the light is visible, a material with the right bandgap must be chosen. Table 12.1 shows the materials listed in Table 11.1 again, with colours indicated – all but two are in the infrared, and the visible two are indirect energy gap materials.

If you want an LED for which the wavelength doesn't matter – for remote control devices and so on – then gallium arsenide is a good material to use. There is a well developed technology for the manufacture and processing of gallium arsenide, and it is an efficient emitter. For many applications the fact that its output is outside the visible range is advantageous.

12.1 VISIBLE-LIGHT EMITTING DIODES

There are some other two-element compound semiconductor materials with

Table 12.1

Semiconductor material	Direct or indirect gap (D or I)	Emission wavelength (μm)	Colour
Indium antimonide (InSb)	D	7.29	
Indium arsenide (InAs)	D	3.44	
Germanium (Ge)	I	1.85	
Gallium antimonide (GaSb)	D	1.72	Infrared
Silicon (Si)	I	1.11	
Indium phosphide (InP)	D	0.92	
Gallium arsenide (GaAs)	D	0.87	
Gallium phosphide (GaP)	I	0.51	Green
Silicon carbide (SiC)	I	0.41	Blue

energy gaps that put their emission wavelengths in the visible region, but the technology required to use them is very difficult, so it seems that all we have are two inefficient emitting materials for green and blue wavelengths and nothing for red.

The indirect energy gap materials can have their radiative carrier lifetimes reduced, and so their internal quantum efficiencies increased, by adding special dopants which trap carriers and so form 'radiative recombination centres'. Such dopants introduce energy levels a little below the conduction band and so move the output light to lower frequencies – that is to longer wavelengths. Adding nitrogen to gallium phosphide proves to be particularly effective and produces a very efficient emitter, but its output is red. Silicon carbide blue LEDs are available, but they are difficult to make and not highly efficient.

LEDs giving light of colours red through orange and yellow to green can be made by using different compositions of the three-element alloy gallium arsenide phosphide. The arsenic and the phosphorus are the trivalent materials and together must equal the amount of gallium. For the red end there has to be more arsenic than phosphorus and the energy gap is direct. As the percentage of phosphorus increases, a point is reached where the energy gap becomes indirect, so that some nitrogen (or similar) doping is needed, but it is still possible to get green light emission. The material can be grown as an epitaxial layer on gallium arsenide or gallium phosphide, depending on the composition, without too much lattice distortion at the interface.

The structure of the diode must be such that the light, once generated, can get out. When emitted, it sets off in all directions, and much of it is reabsorbed, so that the external quantum efficiency – defined as the ratio of the number of photons *emerging from the device* to the number of electron–hole pairs recombining – is much smaller than the internal quantum efficiency, but we don't want the difference to be any greater than necessary. The structure shown in Fig. 12.1 will achieve this. The substrate must be heavily doped so that it does

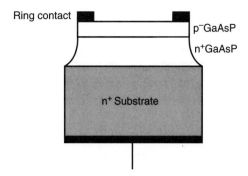

Figure 12.1   A visible LED structure.

not present a significant resistance to the diode current. The epitaxial layer contains a pn junction between heavily doped n-type material and lightly doped p-type material, so that the diode current consists almost entirely of electrons injected into the p side. The light is thus generated in the p material, and the p layer must be wide enough for radiative recombination to occur before the electrons can diffuse to the contact (where they would recombine non-radiatively), but no wider, so that there will be as little reabsorption of light as possible.

12.2   LEDS FOR OPTICAL COMMUNICATIONS

For short-distance links, gallium arsenide diodes can be used with a wavelength of 0.87 μm, but as explained in Chapter 11, for long-distance communications over silica fibre it is necessary to use a wavelength of either 1.3 μm or 1.55 μm, so materials which emit at these two wavelengths must be found.

Indium gallium arsenide, lattice-matched to an indium phosphide substrate, is used in photodiodes to receive at both these wavelengths, but its energy gap corresponds to a radiation wavelength of 1.65 μm, so it will not do as an emitter for either of the required wavelengths. The correct radiating wavelength can be obtained by altering the proportions of indium and gallium, but the resulting material is not lattice-matched to any suitable substrate material. A solution has been found by introducing a fourth component to the alloy – phosphorus. Compositions have been found for indium gallium arsenide phosphide which are lattice-matched to indium phosphide and which radiate at any one of a range of wavelengths covering both the 1.3 μm and the 1.55 μm bands. To keep the composition, plus the doping, exact is tricky, but only a thin layer needs to be laid down.

The favoured structure that has evolved is a double heterostructure. Figure 12.2 shows a layer of n-type indium gallium arsenide phosphide sandwiched between P-type indium phosphide and N-type indium phosphide. When the Pn junction is forward-biased, holes are injected into the n-type material; because

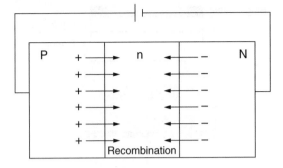

Figure 12.2   A light-emitting diode heterostructure.

of the properties of a heterojunction, very few electrons pass the other way. Electrons enter the n-type material from the N-type layer to maintain charge neutrality. When, however, the holes have diffused as far as the interface with the N-type material they cannot go any further; because of the wider energy gap there are no available energy levels in the N-type material for holes near the top of the Valence band in the n-type material. The carriers are trapped in the narrow band material and must stay there until they recombine, so carrier densities build up until the rate of recombination equals the rate of injection. Since the indium gallium arsenide phosphide is direct energy gap material, a high proportion of recombinations are radiative.

Light can get out through either indium phosphide layer – both are transparent – but another useful idea has been introduced. The indium phosphide layers have a lower refractive index than the indium gallium arsenide phosphide layer, so light which strikes the interface between the centre layer and either sandwiching layer at an angle is liable to suffer total internal reflection.

The active part of the device is made long and thin and one end is made reflective. Most of the light is collected together by reflections so that it emerges from one end. This also makes it easier to couple the light into an optical fibre. The device I have described is a double-heterostructure edge-emitting LED; a sketch of its structure is shown in Fig. 12.3. The wavelength spread of the radiation emitted by such an LED is typically about 60 nm.

[***Note***   Total internal reflection of light occurs when light strikes the interface between the medium in which it is travelling and a medium of lower refractive index provided that it strikes that interface at an angle which is greater than a 'critical angle' – the angle whose sine is the ratio of the refractive indices.]

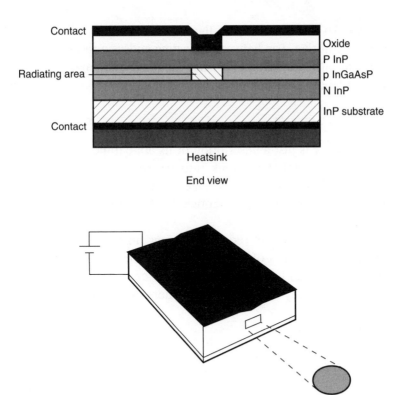

Figure 12.3

12.3 SEMICONDUCTOR LASERS

The light from an LED, like the light from most other sources, is incoherent. This means that its photons are not in phase with each other and the sum of them all is not a wave with a constant phase across a wavefront or a uniform phase lag with distance. Light with the properties I have just described would be called 'spacially and temporally coherent' (like radio waves at lower frequencies) and can only be produced by stimulated emission.

Stimulated emission occurs if an electron is waiting to fall in energy and give out a photon when another photon, of exactly the frequency of the one it is about to produce, comes by and 'stimulates' it to give out its photon; the two photons then go on together in phase.

In normal LED operation, just as there is some absorption there is also some stimulated emission as the light moves through the material to the emitting surface. If reflecting surfaces can be placed so that the light is sent back through the emitting material, making several passes, then stimulated emission can become the dominant emitting mechanism. The device begins to 'lase'.

A favoured design is a double heterostructure, like that of the edge-emitting LED but with both end faces made reflective – the output end has to be only partially reflective so that a fraction of the light can emerge at each reflection. The length of the emitting strip between reflecting ends is a resonant cavity, and reinforcement can only occur at frequencies for which the length of the strip is an exact whole number of half-wavelengths so that it supports a standing wave.

A 1.55 μm oxide stripe laser

The name of this laser is just a description of its appearance, in which the top contact appears as a metal stripe sunk into the oxide protective coating. In structure it is very similar to the double-heterostructure edge-emitting LED, but smaller. The materials used are identical to those used in the LED giving the same wavelength.

For the levels of power output required – a few milliwatts – heat dissipation puts a lower limit on the area of the device, so the resonator length, between reflectors, cannot be much less than about a quarter of a millimetre (250 μm). Let me calculate how many half-wavelengths there are in the standing wave in the cavity.

For N half-wavelengths

$$\frac{N\lambda_m}{2} = L$$

But λ_m is the wavelength in the material, and I only know the free-space wavelength, λ_0. The refractive index of the indium gallium arsenide phosphide for this wavelength is ≈3.5, so taking this figure as exact for the moment,

$\lambda_m = 1.55/3.5 \ \mu$m ≈ $0.443 \ \mu$m

Taking L as 250 μm,

$$N = \frac{2 \times 250 \times 10^{-6}}{0.443 \times 10^{-6}} = 1128.67$$

If the figures used for refractive index and cavity length were exactly accurate, then this answer indicates that there would not be an output at precisely 1.3 μm wavelength.

Using the figures, I shall calculate the exact free-space wavelength associated with 1128 half-wavelengths and 1129 half-wavelengths in the cavity.

For $N = 1128$,

$$\lambda_m = \frac{2 \times 250}{1128} \ \mu m = 0.443 \, 2624 \ \mu m$$

So, since $\lambda_0 = 3.5\lambda_m$, $\lambda_0 = 1.551 \, 418 \ \mu$m. Similarly, for $N = 1129$, $\lambda_0 = 1.550 \, 044 \ \mu$m. The wavelength separation of these results is significant, and is seen to be about 1.38 nm.

These two wavelengths, and others that can be calculated in the same way using values of N near 1128, are 'modes' of the laser. The indium gallium arsenide phosphide itself only produces radiation over a small bandwidth, and furthermore differences in the initial amplitude of emission are increased by the multiple passes of the radiation through the cavity, so the amplitude of the modes falls off rapidly as one moves away from the centre wavelength. The output spectrum of the laser will be similar to Fig. 12.4. About five modes have significant amplitude, and the total wavelength spread is only about 5.5 nm.

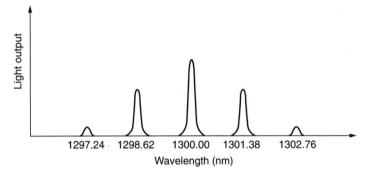

Figure 12.4   A laser spectrum.

Having achieved this spectrum it is possible to eliminate all but the centre mode by using a more complex cavity structure; such lasers are called 'single-mode' lasers and have wavelength spreads of less than 0.01 nm.

If the current through the diode is small, laser action does not take place, and the device acts as a rather inefficient LED. At a certain current, called the 'lasing threshold', stimulated emission takes over and the light output power rises rapidly with the current. A typical characteristic curve is shown in Fig. 12.5. Unfortunately the value of the lasing threshold is very temperature-sensitive and may change from specimen to specimen and with ageing. This means that to maintain a consistent light power output, feedback control systems need to be applied to the laser.

Lasers for other wavelengths

Lasers to operate at 1.3 μm wavelength can be made with the same structure as described above but using the composition of indium gallium arsenide phosphide appropriate for the wavelength. To use a double heterostructure for other wavelengths requires a material with the appropriate bandgap for the active layer and an appropriate material with a wider energy gap and lattice-matched to the active layer for the confining layers. Successful lasers giving a wavelength of 0.87 μm have been made with an active layer of gallium arsenide and confining layers of gallium aluminium arsenide.

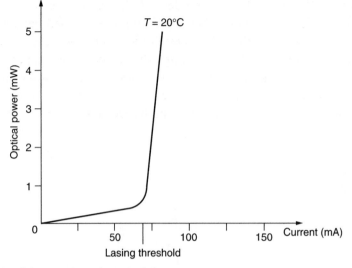

Figure 12.5   A laser output characteristic.

12.4   SUMMARY

When a current flows in a forward-biased pn junction in which one side is much more heavily doped than the other, a large population of excess holes and electrons recombine in the neutral region of the less heavily doped side. If the excess carriers have a short radiative lifetime, then many of the recombinations result in a photon of light, of a wavelength at or near that associated with the energy gap of the material.

Visible LEDs over a colour range from red to green can be made using an epitaxial layer of gallium arsenide phosphide. The proportion of arsenic to phosphorus determines the colour. For compositions giving green light the proportion of phosphorus is such that the energy gap of the material is indirect, and some doping with material to provide radiative recombination centres is necessary. Efficient red LEDs can also be made from gallium phosphide doped with nitrogen. Blue LEDs can be made using silicon carbide.

Light emitters for use in telecommunications, at wavelengths around 1.3 μm and 1.55 μm, are made using indium gallium arsenide phosphide which is lattice-matched to indium phosphide. A favoured structure is the double heterostructure which, when a forward diode current flows, traps the excess carriers until they recombine. The confining layers of indium phosphide can also be used to contain the light produced, by total internal reflection, so that it emerges from one edge of the device in a narrow cone suitable to feed into an optical fibre. LEDs can be made in this form and so can lasers.

A laser produces coherent light. It does this by retaining, by multiple reflections across a cavity, a small number of photons which have been spontaneously released so that they stimulate the emission of many more photons in phase with them. Because of size restrictions, a semiconductor laser has a cavity of length many half-wavelengths, so that within the spectral band of the emitting material several modes of output, associated with slightly different numbers of half-wavelengths across the cavity, are possible. More sophisticated cavity design can eliminate all but one mode, giving a single-mode laser.

A laser has a lasing threshold, which is a current below which it ceases to produce coherent light and behaves like an LED.

12.5 PROBLEMS

1 A gallium arsenide LED has 1.5 V applied and passes 100 mA of current. The internal quantum efficiency is 50 per cent and 1/40 of the light produced is able to emerge from the emitting surface. Calculate the optical power output and the rate of heat dissipation in the diode.

2 Calculate the mode separation in an oxide stripe laser with a cavity length of 250 μm designed for 1.3 μm output. Assume that the refractive index of the indium gallium arsenide phosphide layer is 3.5.

3 From the graph shown in Fig. 12.5, estimate the external quantum efficiency of the device when it is emitting 5 mW of light of wavelength 1.55 μm. Given that the voltage across the device at 5 mW output is 1.2 V, deduce the conversion efficiency of electrical power to optical power.

Solutions to problems

CHAPTER 2

1. If a voltage, V, is connected across the two sheets then one of them will gain a charge $+Q$ and the other $-Q$. Equal and opposite charges will be induced on the faces of the metal block. The flux density in the gaps (assuming fringing flux is negligible) will be

$$Q/A = Q/10^{-4} = 10^4 Q \text{ C m}^{-2}$$

Hence the electric field in the gaps will be D/ϵ_0.

$$E = 10^4 Q/8.85 \times 10^{-12} = 1.13 \times 10^{15} Q \text{ V m}^{-1}$$

There is no electric field in the metal.

The total voltage between the plates is the product of the electric field in the gaps and the total gap length:

$$V = 1.13 \times 10^{15} Q \times 5 \times 10^{-4} = 5.65 \times 10^{11} Q \text{ volts}$$

$$C = Q/V = 1/5.65 \times 10^{11} = \textbf{1.77 pF}$$

2. The current density in the wires is

$$I/A = 1/10^{-6} = 10^6 \text{ A m}^{-2}$$

Since $J = \sigma E$, the electric field in the two wires are as follows: in the lower conductivity wire $10^6/2 \times 10^4 = 50 \text{ V m}^{-1}$; and in the higher conductivity wire $10^6/4 \times 10^4 = 25 \text{ V m}^{-1}$.

Assume that the current is flowing from the lower conductivity wire into the higher. Applying Gauss's theorem to the junction, the electric flux into the junction is

$$AD = A\epsilon E$$

$$\epsilon = \epsilon_0\epsilon_r = 8.85 \times 10^{-12} \times 5.5 = 4.87 \times 10^{-11} \text{ F m}^{-1}$$

$$\text{flux}_{\text{in}} = 10^{-6} \times 4.87 \times 10^{-11} \times 50 = 2.44 \times 10^{-15} \text{ C}$$

Similarly $\text{flux}_{\text{out}} = 10^{-6} \times 4.87 \times 10^{11} \times 25 = 1.22 \times 10^{15}$ C. The charge at the junction must be the difference between these, i.e. a negative charge of $\textbf{1.22} \times \textbf{10}^{-15}$ **C**

If the current were flowing the other way, the electric flux out would be greater than the flux in, so the stored charge would have the same value but positive.

The number of electronic charges stored is

$$1.22 \times 10^{-15}/1.6 \times 10^{-19} = \textbf{7625}$$

3. The conductivity of $1/10^{-8} = 10^8$ S m^{-1}

   $\sigma = ne\mu$

   and n for copper $\approx 10^{29}$ m^{-3}. This yields $\mu = \sigma/ne = 10^8/(10^{29} \times 1.6 \times 10^{-19})$
   $= \mathbf{6.25 \times 10^{-3} \ m^2 \ V^{-1} \ s^{-1}}$

4. $\sigma = ne\mu = 10^{23} \times 1.6 \times 10^{-19} \times 0.15 = \mathbf{2.4 \times 10^3 \ S \ m^{-1}}$.

   $J = nev_d$

   $J = I/A = 100 \ \text{mA}/1\text{mm}^2 = 0.1/10^{-6} = 10^5$ A m^{-2}. So

   $v_d = J/ne = 10^5/(10^{23} \times 1.6 \times 10^{-19}) = \mathbf{6.25 \ m \ s^{-1}}$

CHAPTER 3

1. The density of states constant is

$$N_c = \frac{n_i}{\exp(-E_g/2kT)}$$

At room temperature, $2kT = 0.05$ eV. Taking values of n_i and E_g from Table 1.2,

Silicon, $N_c = 10^{16}/\exp(-1.12/0.05) \approx \mathbf{5 \times 10^{25} \ m^{-3}}$

Germanium, $N_c = 10^{19}/\exp(-0.67/0.05) \approx \mathbf{7 \times 10^{24} \ m^{-3}}$

Gallium arsenide, $N_c = 10^{12}/\exp(-1.42/0.05) \approx \mathbf{2 \times 10^{24} \ m^{-3}}$

Indium phosphide, $N_c = 10^{14}/\exp(-1.35/0.05) \approx \mathbf{5 \times 10^{25} \ m^{-3}}$

All the materials have similar crystal structures and so should have broadly similar band structures. There will be differences of inter-atomic distance in the different crystals, so there will not be the same number of atoms per cubic metre – which will affect the value of N_c.

2. As shown in the text,

$$\frac{n_i(291)}{n_i(290)} = \frac{\exp(-E_g/582k)}{\exp(-E_g/580k)} = \exp\left[\left(\frac{E_g}{k}\right)\left(\frac{1}{580} - \frac{1}{582}\right)\right]$$

$1/k(580^{-1} - 582^{-1}) = 0.069(\text{eV})^{-1}$

So a 1° rise above room temperature causes an increase of $\exp(0.069E_g)$.

   Hence the material with the most rapid rise of n_i with temperature will be the one with the largest E_g, i.e. **gallium arsenide**.

$\exp(0.069 \times 1.42) = 1.103 =$ a rise of **10.3%**

3. $\sigma = n_i e(\mu_n + \mu_p) = 2n_i e\mu$ (in this case)

$$\sigma = 1/\text{resistivity} \approx 10^{-14} \ \text{S m}^{-1}$$

$$2n_i e\mu = 10^{-14}, \text{ so } n_i \approx 10^{-14}/(2 \times 1.6 \times 10^{-19} \times 0.01)$$

$$= 3 \times 10^6, \text{ order of magnitude of } n_i \text{ is } \mathbf{10^6 \ m^{-3}}$$

$$\exp(-E_g/2kT) = n_i/N_c \approx 10^6/10^{25} = 10^{-19}$$

$$E_g/2kT = -\ln 10^{-19} \approx 44$$

At room temperature $2kT = 0.05$ eV. Thus

$$E_g \approx 44 \times 0.05 = \textbf{2.2 eV}$$

4. The probability of occupancy of an energy level at the bottom of the conduction band is given by

$$f(E_B) \approx 1/\exp[(E_B - E_F)/kT]$$

Similarly, at the top of the band,

$$f(E_T) \approx 1/\exp[(E_T - E_F)/kT]$$

The ratio of these probabilities,

$$f(E_T)/f(E_B) = \frac{\exp[(E_B - E_F)/kT]}{\exp[(E_T - E_F)/kT]}$$

$$= \exp[(E_B - E_T)/kT]$$

The difference in energy between the top and bottom of the conduction band is given in the text and in Fig. 3.2 as 3.4 eV, so the ratio of probabilities is

$$\exp(-3.4/0.025) = \textbf{8} \times \textbf{10}^{-60}$$

The number of electrons in 1 cm³ of silicon at room temperature $= 10^{16}/10^6 = 10^{10}$. Most of these will be near the bottom of the conduction band, so taking this as the effective number at the bottom, the number at the top is $8 \times 10^{-60} \times 10^{10} = 8 \times 10^{-50}$, i.e. none.

No electrons have enough energy to escape.

CHAPTER 4

1. (a) 1 mm³ $= 10^{-9}$ m³, so there are 4×10^{19} germanium atoms in 1 mm³, and hence

$$4 \times 10^{19}/10^5 = \textbf{4} \times \textbf{10}^{14} \textbf{ aluminium atoms}$$

   (b) Aluminium is an acceptor impurity, so holes are the majority carriers and, effectively, the number of majority carriers is $\textbf{4} \times \textbf{10}^{14}$ **holes**.
   From Table 1.2, n_i for Ge at room temperature $\approx 10^{19}$, so in the specimen there are $10^{19} \times 10^{-9} = 10^{10}$ intrinsic carriers, so

$$\text{the number of minority carriers} = (10^{10})^2/4 \times 10^{14} = \textbf{2.5} \times \textbf{10}^5 \textbf{ electrons}$$

   (c) The temperature limit for extrinsic behaviour is taken as the temperature where

$$n_i = 4 \times 10^{28}/10^5 = 4 \times 10^{23} \text{ m}^{-3}$$

$$n_i = N_c\exp(-E_g/2kT)$$

and N_c for germanium has been calculated in the solution of question one of Chapter 3, to be $\approx 7 \times 10^{24}$ m^{-3}. Thus

$$\frac{4 \times 10^{23}}{7 \times 10^{24}} = \exp\frac{(-0.67)}{2 \times 8.62 \times 10^{-5}\, T}$$

$$0.057 = \exp(-3886/T)$$

$$\frac{3886}{T} = -\ln 0.057 = 2.86$$

$$T = 3886/2.86 \approx \textbf{1359 K}$$

Since at this temperature $n_i = N_A$, and since $pn = n_i^2$, I can write

$$p = N_A + n$$

and so

$$(N_A + n)n = N_A^2$$

This yields the quadratic equation

$$n^2 + N_A n - N_A^2 = 0$$

The standard solution is

$$n = \frac{-N_A + \sqrt{(N_A^2 + 4N_A^2)}}{2}$$

$$= \frac{N_A(\sqrt{5} - 1)}{2} = 0.62\, N_A$$

So, $n = 0.62 \times 4 \times 10^{23} \approx 2.5 \times 10^{23}$ m^{-3} and $p = 6.5 \times 10^{23}$ m^{-3}. In 1 mm^3, there are **2.5 $\times$ 10^{14} electrons and 6.5 $\times$ 10^{14} holes**.

2. Phosphorus is a donor impurity, so most of the carriers will be electrons. Hence

$$\sigma \approx ne\mu_n \approx N_D e\mu_n$$

So we require

$$N_D = \sigma/e\mu_n$$

For resistance R, $R = L/\sigma A$, so $\sigma = L/RA$. Here,

$$\sigma = \frac{100 \times 10^{-6}}{10 \times 15 \times 10^{-6} \times 5 \times 10^{-6}}$$

$$= 1.33 \times 10^5 \text{ S m}^{-1}$$

$$N_D = \frac{1.33 \times 10^5}{1.6 \times 10^{-19} \times 0.15}$$

$$= 5.6 \times 10^{24} \text{ m}^{-3}$$

Since there are 5×10^{28} silicon atoms m^{-3}, the required doping density is

$$\frac{5.6 \times 10^{24} \times 10^6}{5 \times 10^{28}} \text{ ppm} = \textbf{112 ppm}$$

3. Arsenic is a donor impurity, so the majority carriers are electrons. Diffusion will carry the electrons away from the more heavily doped end leaving it positive, so the internal electric field will be from the higher doped to the lower doped face.

The potential difference between the faces,

$$V_{ab} = (kT/e) \ln (n_a/n_b) \approx (kT/e) \ln (N_{Da}/N_{Db})$$

$$= 0.025 \ln 10^5 = \textbf{0.29 V}$$

CHAPTER 5

1. $\psi = (kT/e) \ln[n_i^2/(N_A N_D)]$

$$= 0.025 \ln[(10^{19})^2/(10^{22} \times 10^{23})]$$

$$= \textbf{-0.40 V}$$

2. The aluminium doped region is p-type. Thus

$$L_n = \sqrt{(D_n t_n)}$$

$$= \sqrt{(0.01 \times 10^{-7})}$$

$$\approx \textbf{32 } \mu\textbf{m}$$

The arsenic doped region is n-type. Thus

$$L_p = \sqrt{(D_p t_p)}$$

$$= \sqrt{(0.0049 \times 10^{-7})}$$

$$\approx \textbf{22 } \mu\textbf{m}$$

3. $w_t = \sqrt{\left[\frac{2\epsilon_0 \epsilon_r}{e}\left(\frac{1}{N_A} + \frac{1}{N_D}\right)(|\psi| - V_D)\right]}$

$$w_t = \sqrt{(2 \times 8.85 \times 10^{-12} \times 16/1.6 \times 10^{-19})(10^{-22} + 10^{-23})(0.4 - 0.2)} = \textbf{0.197 } \mu\textbf{m}$$

10/11 of this is in the lightly doped p material, so the width of the neutral region in the aluminium doped region is

$$5 \text{ } \mu\text{m} - (10/11) \times 0.197 \text{ } \mu\text{m}$$

$$\approx \textbf{4.8 } \mu\textbf{m}$$

4. In this case, since $l_p \ll L_n$ but $l_n \gg L_p$,

$$I_s = Ae\left(\frac{D_p p_{n0}}{L_p} + \frac{D_n n_{p0}}{l_p}\right)$$

Also,

$$p_{n0} = n_i^2/N_D = (10^{19})^2/10^{23} = 10^{15}$$

and

$$n_{p0} = n_i^2/N_A = (10^{19})^2/10^{22} = 10^{16}$$

So

$$I_s = 10^{-6} \times 1.6 \times 10^{-19} \times [(0.0049 \times 10^{15}/22 \times 10^{-6}) + (0.01 \times 10^{16}/4.8 \times 10^{-6})]$$

$$I_s = 3.4\ \mu A$$

$$I_D = I_s\left[\exp\left(\frac{eV_D}{kT}\right) - 1\right]$$

$$= 3.4 \times 10^{-6}[\exp(0.2/0.025) - 1]$$

$$I_D = 10.1\ mA$$

5. In this case

$$\frac{I_p}{I_n} = \frac{D_p}{D_n} \times \frac{N_A}{N_D} \times \frac{l_p}{L_p}$$

$$= \frac{0.0049 \times 10^{22} \times 4.8 \times 10^{-6}}{0.01 \times 10^{23} \times 22 \times 10^{-6}}$$

$$= 0.0107$$

$$I_n{:}I_p = 0.0107^{-1} = 93.5{:}1$$

6. Since the p side is much shorter than the minority carrier diffusion length,

$$C_{dp} = \frac{l_p^2 I_n}{2D_n} \frac{e}{kT}$$

$$= \frac{(4.8 \times 10^{-6})^2 \times 10.1 \times 10^{-3} \times 93.5/94.5}{2 \times 0.01 \times 0.025} = 460\ pF$$

For the n material

$$C_{dn} = (t_p I_p) \frac{e}{kT}$$

$$= \frac{10^{-7} \times 10.1 \times 10^{-3}/94.5}{0.025} = 428\ pF$$

The total diffusion capacitance is $\approx$ **890 pF**

7. When $V_D = -2$ V,

$$w_t = \sqrt{\left[\frac{2\epsilon_0\epsilon_r}{e}\left(\frac{1}{N_A} + \frac{1}{N_D}\right)(|\psi| - V_D)\right]}$$

$$w_t = \sqrt{[(2 \times 8.85 \times 10^{-12} \times 16/1.6 \times 10^{-19})(10^{-22} + 10^{-23})(0.4 + 2)]}$$

$$= \textbf{0.684 } \boldsymbol{\mu}\textbf{m}$$

The amount of the transition region in the p material is now

$(10/11) \times 0.684~\mu m = 0.62~\mu m$

so the width of the neutral p region is now $\approx 4.4~\mu m$. So

$$I_s = 10^{-6} \times 1.6 \times 10^{-19} \times [(0.0049 \times 10^{15}/22 \times 10^{-6}) + (0.01 \times 10^{16}/4.4 \times 10^{-6})]$$

$$I_s = \textbf{3.7 } \boldsymbol{\mu}\textbf{A}$$

The increase is due to the narrowing of the p-side neutral region.

8. $C_t = A\sqrt{\left[\dfrac{e\epsilon_0\epsilon_r}{2(1/N_A + 1/N_D)(|\psi| - V_D)}\right]}$

$$= 10^{-6}\sqrt{(1.6 \times 10^{-19} \times 8.85 \times 10^{-12} \times 16)/[2(10^{-22} + 10^{-23})(0.4 + 2)]}$$

$$\approx \textbf{200 pF}$$

CHAPTER 6

1. (a)   From Chapter 5,

$$E_{max} = \frac{2|V_D + \psi|}{w_t}$$

and

$$w_t = \left[\frac{2\epsilon_0\epsilon_r}{e}\left(\frac{1}{N_A} + \frac{1}{N_D}\right)(|\psi| - V_D)\right]$$

V_D is negative here, and should be $\gg |\psi|$, so, approximately,

$$E_{max} = \sqrt{\left(\frac{2V_D e}{\epsilon_0\epsilon_r(N_A^{-1} + N_D^{-1})}\right)}$$

giving

$$V_D = E_{max}^2\epsilon_0\epsilon_r(N_A^{-1} + N_D^{-1})/2e$$

Assuming that breakdown occurs when $E_{max} = 5 \times 10^7$ V m^{-1},

$$V_{BD} = (5 \times 10^7)^2 \times 8.85 \times 10^{-12} \times 12 \times 2 \times 10^{-22}/2 \times 1.6 \times 10^{-19}$$

$$\approx \textbf{165 V}$$

(b) I can get I_s for the junction, in this case, from

$$I_s = Ae\left[\frac{D_p p_{n0}}{L_p} + \frac{D_n n_{p0}}{L_n}\right]$$

Now, $p_{n0} = n_i^2/N_D = 10^{32}/10^{22} = 10^{10}$ m^{-3}. Similarly $n_{p0} = 10^{10}$ m^{-3}. Thus

$$I_s = 5 \times 10^{-6} \times 1.6 \times 10^{-19} \times (0.0012 \times 10^{10}/15 \times 10^{-6} + 0.0039 \\ \times 10^{10}/30 \times 10^{-6})$$

$$= 1.7 \times 10^{-12} \text{ A}$$

Since

$$I_D \approx I_s \exp(eV_D/kT)$$

$$V_D = (kT/e)\ln(I_D/I_s)$$

$$= 0.025 \ln(1/1.7 \times 10^{-12}) = 0.68 \text{ V}$$

I must add to this the resistances of 70 μm of p material and 85 μm of n material.
For the p material,

$$\sigma \approx N_A e\mu_p = 10^{22} \times 1.6 \times 10^{-19} \times 0.045$$

$$= 72 \text{ S m}^{-1}$$

$$R = L/\sigma A = 70 \times 10^{-6}/72 \times 5 \times 10^{-6} = 0.19 \text{ }\Omega$$

The voltage dropped across this by 1 A = 0.19 V.
For the n material,

$$\sigma \approx N_D e\mu_n = 10^{22} \times 1.6 \times 10^{-19} \times 0.15$$

$$= 240 \text{ S m}^{-1}$$

$$R = L/\sigma A = 85 \times 10^{-6}/240 \times 5 \times 10^{-6} = 0.07 \text{ }\Omega$$

The voltage dropped across this by 1 A = 0.07 V.
The total voltage across the diode is 0.68 + 0.19 + 0.07 V = **0.94 V**.

2. (a) For $V_D = 0.6$ V,

$$\exp(eV_D/kT) = \exp(0.6/0.025)$$

$$= 2.65 \times 10^{10}$$

In the diode equation $I_D = 5 \times 10^{-14} \times 2.65 \times 10^{10} = 1.3$ mA.

For $V_D = 0.7$ V,

$$\exp(eV_D/kT) = \exp(0.7/0.025)$$

$$= 1.45 \times 10^{12}$$

In the diode equation $I_D = 5 \times 10^{-14} \times 1.45 \times 10^{12} = 72.3$ mA.
The range of currents is from 1.3 to 72.3 mA.

(b) $r_e = 25/(I_D \text{ mA}) = $ **19 Ω to 0.35 Ω**

(c)   Assuming that all the diffusion capacitance is due to charge in the p side (because of the injection ratio), L_n in the p material $= \sqrt{(D_n t_n)} = \sqrt{(0.0039 \times 2 \times 10^{-7})} = 28\ \mu m$, which is $<l_p$, so the appropriate equation is

$$C_{dp} = (t_n I_n)e/kT$$

Thus

$$C_d \approx t_n \times (e/kT) \times I_D$$

giving, for $I_D = 1.3$ mA, $C_d \approx$ **10.4 nF**; and for $I_D = 72.3$ mA, $C_d \approx$ **578 nF**.

CHAPTER 8

1. $$V_D = \frac{\left|\dfrac{Q_D}{A}\right| + \left|\dfrac{Q_{SS}}{A}\right|}{C_{ox}} + |\psi_D|$$

$$\psi_D = \frac{2kT}{e}\ln\left(\frac{N_{A/D}}{n_i}\right)$$

So in this case, $\psi_D = 0.05\ln(10^{23}/10^{16}) = 0.81$ V. Now

$$\frac{Q_D}{A} = \pm\sqrt{(2\epsilon_0\epsilon_r e\psi_D N_{A/D})}$$

$$= \sqrt{(2 \times 8.85 \times 10^{-12} \times 12 \times 1.6 \times 10^{-19} \times 0.81 \times 10^{23})}$$

$$= 1.66 \times 10^{-3}\ \mathrm{C\,m}^{-2}$$

$$\frac{Q_{SS}}{A} = e(N_{ox} \pm N_{sd})$$

The surface layer of phosphorus atoms, being donors, are positive ions, so

$$Q_{SS}/A = 1.6 \times 10^{-19}(10^{15} + 1.5 \times 10^{16}) = 2.56 \times 10^{-3}\ \mathrm{C\,m}^{-2}$$

$$C_{ox} = \frac{4\epsilon_0}{t_{ox}} = 4 \times 8.85 \times 10^{-12}/10^{-7} = 3.5 \times 10^{-4}\ \mathrm{F\,m}^{-2}$$

The boron atoms, being acceptors, form negative ions, and Q_{SS} has come out positive, so

$$V_T = \frac{1.66 \times 10^{-3} - 2.56 \times 10^{-3}}{3.5 \times 10^{-4}} + 0.81 = -\textbf{1.76 V}$$

This will be an n-channel depletion mode device.

2. $V_{GS} - V_T = 6.76$ V and $V_{DS} = 10$ V, so the transistor is in the saturated region of operation.

$$I_D = \tfrac{1}{2}\beta(V_{GS} - V_T)^2[1 + \lambda(V_{DS} + V_T - V_{GS})]$$

and

$$\beta = \frac{W\mu C_{OX}}{L}$$

μ being taken as half the normal value for electrons in silicon.

$\beta = \quad 15 \times 10^{-6} \times 0.075 \times 3.5 \times 10^{-4}/10 \times 10^{-6}$

$\quad = \quad 3.9 \times 10^{-5}$

$I_D = \quad 0.5 \times 3.9 \times 10^{-5} \times 6.76^2[1 + 0.01(10 - 6.76)]$

$\quad = \quad \textbf{0.92 mA}$

3. In the saturated region,

$g_m \approx \sqrt{(2\beta I_D)}$

which is the case here, so

$g_m = \sqrt{(2 \times 3.9 \times 10^{-5} \times 0.92 \times 10^{-3})}$

$\quad = \textbf{268 } \boldsymbol{\mu}\textbf{A V}^{-1}$

$g_{ds} \approx \dfrac{I_D}{V_{DS} + 1/\lambda}$

$\quad = 0.92 \times 10^{-3}/(10 + 0.01^{-1})$

$\quad = \textbf{8.4} \times \textbf{10}^{-6}\textbf{ S}$

4. C_{gs} and C_{gd} are each approximately half the gate–channel capacitance, which is

$C_{ox} \times W \times L = 3.5 \times 10^{-4} \times 15 \times 10^{-6} \times 10 \times 10^{-6}$

$\qquad\qquad = 5.25 \times 10^{-14} \text{ F}$

So, $C_{gs} = C_{gd} \approx \textbf{2.6} \times \textbf{10}^{-14}\textbf{ F}$.

CHAPTER 9

1. The collector currents are different, since the diffusion constant of holes in the base is smaller than the diffusion constant of electrons.

For $V_{CE} = 2$ V,

$I_C = AeD_p p_{n1}/W$

$\quad = 5 \times 10^{-8} \times 1.6 \times 10^{-19} \times 0.0012 \times 2.6 \times 10^{20}/2.67 \times 10^{-6}$

$\quad = 0.93 \text{ mA}$

For $V_{CE} = 12$ V,

$I_C = 5 \times 10^{-8} \times 1.6 \times 10^{-19} \times 0.0012 \times 2.6 \times 10^{20}/2.45 \times 10^{-6}$

$\quad = 1.02 \text{ mA}$

So $g_o = (1.02 \times 10^{-3} - 0.93 \times 10^{-3})/10 = 9 \times 10^{-6}$ S, i.e. $g_o = \textbf{9 } \boldsymbol{\mu}\textbf{S}$.

For the emitter–base junction

$$\frac{I_p}{I_n} = \frac{D_p}{D_n} \times \frac{N_A}{N_D} \times \frac{\text{width of emitter neutral region}}{\text{active base width}}$$

Taking a representative value of W as 2.5 μm,

$$I_p/I_n = 0.0012 \times 10^{24} \times 5 \times 10^{-6}/0.0039 \times 10^{22} \times 2.5 \times 10^{-6}$$

$$= 62, \text{ giving } \beta = \mathbf{62}$$

Taking the current as nominally 1 mA, $r_e \approx \mathbf{25 \ \Omega}$

$$r_{b'e} = \beta r_e = 62 \times 25, r_{b'e} = \mathbf{1550 \ \Omega}$$

$$g_m = 1/r_e, g_m = \mathbf{40 \ mA \ V^{-1}}$$

The transition region capacitances will be the same: for the base diffusion capacitance

$$C_D \approx \frac{W^2 I_C}{2D_p} \frac{e}{kT}$$

which is also unchanged, because I_C is reduced in proportion to the smaller diffusion constant of holes.

Using

$$g_o = I_C/(V_{CE} + VA)$$

$$9 \times 10^{-6} = 1.02 \times 10^{-3}/(12 + VA)$$

$$12 + VA = 113$$

$$VA = \mathbf{101}, \text{ effectively unchanged}$$

Since VA is unchanged this indicates that the reduction in g_o is due simply to the lower collector current for the applied base voltage. The only significant difference between the pnp and the npn transistor is that the pnp device has a smaller value of β.

2. It is first necessary to calculate the active base width. For this I need values of ψ.

$$\psi = (kT/e) \ln[n_i^2/(N_A N_D)]$$

For the emitter–base junction

$$\psi = 0.025 \ln(10^{32}/5 \times 10^{46})$$

$$= -0.85 \ V$$

For the base–collector junction

$$\psi = 0.025 \ln(10^{32}/5 \times 10^{43})$$

$$= -0.67 \ V$$

$$w_t = \sqrt{\left[\frac{2\epsilon_0 \epsilon_r}{e} \left(\frac{1}{N_A} + \frac{1}{N_D} \right)(|\psi| - V_D) \right]}$$

For the base–emitter junction

$$w_t = \sqrt{[2 \times 8.85 \times 10^{-12} \times 12(2 \times 10^{-25} + 10^{-22})(0.85 - 0.6)/1.6 \times 10^{-19}]}$$

$$= 0.18 \, \mu m, \text{ effectively all in the base}$$

For the base–collector junction

$$w_t = \sqrt{[2 \times 8.85 \times 10^{-12} \times 12(2 \times 10^{-22} + 10^{-22})(0.67 + 5 - 0.6)/1.6 \times 10^{-19}]}$$

$$= 1.42 \, \mu m, \text{ one third in the base, i.e. } 0.47 \, \mu m$$

Active base width, $W = 4 - 0.18 - 0.47 = 3.35 \, \mu m$. Injection ratio,

$$\frac{I_p}{I_n} = \frac{D_p}{D_n} \times \frac{N_A}{N_D} \times \frac{\text{width of emitter neutral region}}{\text{active base width}}$$

$$= 0.0012 \times 5 \times 10^{24} \times 6 \times 10^{-6}/0.0039 \times 10^{22} \times 3.35 \times 10^{-6}$$

$$= 276$$

Call the hole current at the emitter I_{in}. The electron current into the emitter is $I_{in}/277$, so this must be the value of part of the base current. Let the hole current lost by recombination in the base be δI. Now,

$$I_{in} - \delta I = 990 \times 10^{-6}$$

Also,

$$I_{in}/277 + \delta I = 4.5 \times 10^{-6}$$

Adding,

$$I_{in}/277 + I_{in} = 994.5 \times 10^{-6}$$

$$I_{in} = 994.5 \times 10^{-6}/(1 + 1/277) = 990.9 \times 10^{-6} \, A$$

The percentage of holes that reach the collector is $(990/990.9) \times 100 = \textbf{99.91\%}$.

CHAPTER 10

1. From the graph, the absorption edge for the InGaAsP is at $1.3 \, \mu m$. Its energy gap is

$$hc/\lambda_0 = 4.14 \times 10^{-15} \times 3 \times 10^8/1.3 \times 10^{-6} \, eV$$

$$= \textbf{0.96 eV}$$

Similarly, for the InGaAs,

$$E_g = 4.14 \times 10^{-15} \times 3 \times 10^8/1.65 \times 10^{-6} \, eV$$

$$= \textbf{0.75 eV}$$

2. The number of photons surviving to a distance of x is

$$N_x = N_0 \exp(-\alpha x)$$

so, when $N_x/N_0 = 0.05$ (5%), $-\alpha x = \ln 0.05 = -3$, and

$$x = 3/5 \times 10^5 = \textbf{6 } \boldsymbol{\mu m}$$

3. For any wave, $v = f\lambda$; here $v = c/n = 3 \times 10^8/3.5 \text{ ms}^{-1}$. Thus

$f = v/\lambda = 3 \times 10^8/(3.5 \times 0.4 \times 10^{-6}) = \mathbf{2.14 \times 10^{14} \text{ Hz}}$

4. $\quad n_{\text{int}} = \dfrac{1/t_{\text{rr}}}{1/t_{\text{rr}} + 1/t_{\text{nr}}}$

$\quad\quad = 10^7/(10^7 + 5 \times 10^6)$

$\quad\quad = \mathbf{0.67}$

CHAPTER 11

1. Look at Figs 11.2 and 11.3. The diode forward-biases itself, but the net current is a reverse current. The total power flux on the cell is $10^3 \times 10^{-4} \text{ watts} = 100 \text{ mJ s}^{-1}$.
    At the average wavelength,

$E_{\text{ph}} = hc/\lambda_0$

$\quad\quad = 6.63 \times 10^{-34} \times 3 \times 10^8/0.7 \times 10^{-6} \text{ J} = 2.84 \times 10^{-19} \text{ J}$

So, the photon flux is $0.1/2.84 \times 10^{-19} \text{ s}^{-1} = 3.52 \times 10^{17} \text{ s}^{-1}$.
    Half of these are converted to electricity, so the photocurrent consists of 1.76×10^{17} electronic charges per second:

$I_{\text{p}} = 1.76 \times 10^{17} \times 1.6 \times 10^{-19} \approx 28 \text{ mA}.$

(a) $\quad I_{\text{D}} = I_{\text{s}}[\exp(eV_{\text{D}}/kT) - 1] - I_{\text{p}}$

$\quad\quad\quad = 10^{-9}[\exp(0.38/0.025) - 1] - 28 \times 10^{-3}$

$\quad\quad\quad = \mathbf{(-)24 \text{ mA}}$

(b) The voltage across the load is the same as across the diode, so the required R is $0.38/0.24 \approx \mathbf{16 \ \Omega}$

(c) Power delivered to the load $= VI = 0.38 \times 0.024 \approx 9 \text{ mW}$. The light power in $= 100 \text{ mW}$, so the conversion efficiency is **9%**.

2. The photon energy for $1.3 \ \mu\text{m}$ radiation is

$E_{\text{ph}} = hc/\lambda_0 = 6.63 \times 10^{-34} \times 3 \times 10^8/1.3 \times 10^{-6}$

$\quad\quad = 1.53 \times 10^{-19} \text{ J}$

The number of photons per second in $1 \ \mu\text{W} = 10^{-6}/1.53 \times 10^{-19} = 6.54 \times 10^{12}$ photons s^{-1}. A total of 67 per cent of these create electron–hole pairs, i.e. $0.67 \times 6.54 \times 10^{12} = 4.38 \times 10^{12} \text{ s}^{-1}$. The resulting current is $4.38 \times 10^{12} \times 1.6 \times 10^{-19} \text{ A} = 0.7 \ \mu\text{A}$.

(a) In a $1 \ \text{M}\Omega$ resistor current pulses of this amplitude give voltage pulses of **0.7 V**.

(b) When the light is off, the current is the dark current only, which can be assumed to be negligible, so negligible power is dissipated in the diode.
    During light pulses, the voltage across the diode is

$10 - 0.7 = 9.3 \text{ V}$

and the current in it is 0.7 μA. So the electrical power dissipated is
$9.3 \times 0.7 \times 10^{-6} = 6.5\ \mu$W.

In addition, 1 μW of light enters the diode, so, the total power dissipated in the diode during pulses is **7.5 μW**. Notice that in the photodiode the electrical energy in the output signal (across the resistor) does not come from the light; the light simply releases the supply current.

CHAPTER 12

1. 100 mA of current represents $0.1/1.6 \times 10^{-19} = 6.25 \times 10^{17}$ injected electronic charges per second. Of these, half result in photons and 1/40 of these get out, so

Number of emerging photons s^{-1} = $6.25 \times 10^{17}/80$

$$= 7.8 \times 10^{15}\ \text{s}^{-1}$$

The photon energy is equal to the energy gap in GaAs which is
$1.42\ \text{eV} = 1.42 \times 1.6 \times 10^{-19} = 2.27 \times 10^{-19}$ J per photon.

Light power output = $7.8 \times 10^{15} \times 2.27 \times 10^{-19} = $ **1.8 mW**. Total power into the diode = $1.5\ \text{V} \times 0.1\ \text{A} = 150$ mW. Of this $150 - 1.8 = $ **148.2 mW must be dissipated as heat**.

2. For N half-wavelengths,

$$\frac{N\lambda_\text{m}}{2} = L$$

If λ_0 is exactly 1.3 μm,

$\lambda_\text{m} = 1.3/3.5\ \mu\text{m} \approx 0.3714\ \mu$m

So, in 250 μm,

$$N = \frac{2 \times 250 \times 10^{-6}}{0.3714 \times 10^{-6}} = 1346.26$$

I shall calculate the exact free-space wavelengths associated with 1346 half-wavelengths and 1347 half-wavelengths in the cavity. For $N = 1346$,

$$\lambda_\text{m} = \frac{2 \times 250\ \mu\text{m}}{1346} = 0.371\ 4710\ \mu\text{m}$$

and for this, $\lambda_0 = 3.5\lambda_\text{m}$; thus $\lambda_0 = 1.300\ 1486\ \mu$m.
For $N = 1347$,

$$\lambda_\text{m} = \frac{2 \times 250\ \mu\text{m}}{1347} = 0.371\ 1952\ \mu\text{m}$$

and for this, $\lambda_0 = 3.5\lambda_\text{m}$; thus $\lambda_0 = 1.299\ 1834\ \mu$m.
The mode separation is $1.300\ 1486 - 1.299\ 1834\ \mu\text{m} = $ **0.965 nm**.

3. From the graph, the current at 5 mW output is $\approx$ 80 mA. This represents the passage of $0.08/1.6 \times 10^{-19} = 5 \times 10^{17}$ electronic charges s^{-1}.

One photon of 1.55 μm radiation has energy

$$hc/\lambda_0 = 6.63 \times 10^{-34} \times 3 \times 10^8/1.55 \times 10^{-6} = 1.28 \times 10^{-19} \, J$$

So 5 mW of radiation consists of $5 \times 10^{-3}/1.28 \times 10^{-19} = 3.9 \times 10^{16}$ photons s^{-1}.

External quantum efficiency $= 3.9 \times 10^{16}/5 \times 10^{17} = \textbf{7.8\%}$. The input of electrical power $= 1.2 \, V \times 0.08 \, A = 96$ mW. So the conversion efficiency $= 5/96 = \textbf{5.2\%}$.

Summary of formulae

In any semiconductor

$n_i = N_C \exp(-E_g/2kT)$ $N_C \approx 10^{25}$ within an order of magnitude

In equilibrium

$pn = n_i^2$

$n_{n0} \approx N_D, p_{n0} \approx n_i^2/N_D$

$p_{p0} \approx N_A, n_{p0} \approx n_i^2/N_A$

Conductivity contributions

holes $\sigma_p = pe\mu_p$, electrons $\sigma_n = ne\mu_n$

Current density contributions, ignoring signs

Drift: holes $J_{p(dr)} = \sigma_p E = pe\mu_p E$

                electrons $J_{n(dr)} = \sigma_n E = ne\mu_n E$

Diffusion: holes $J_{p(dif)} = eD_p \dfrac{dp}{dx}$

                electrons $J_{n(dif)} = eD_n \dfrac{dn}{dx}$

Equilibrium voltage between two points with different carrier densities

$V_{ab} = (kT/e) \ln (p_b/p_a) = (kT/e) \ln (n_a/n_b)$

For a pn junction

'Diode' equation

$I_D = I_s[\exp(eV_D/kT) - 1]$

Saturation current

$$I_s = Ae\left[\frac{D_p p_{n0}}{(l_n \text{ or } L_p)} + \frac{D_n n_{p0}}{(l_p \text{ or } L_n)} \right]$$

Diffusion lengths

$$L_p = \sqrt{(D_p t_p)} \qquad L_n = \sqrt{(D_n t_n)}$$

Current ratio

$$\frac{I_p}{I_n} = \frac{D_p}{D_n} \times \frac{N_A}{N_D} \times \frac{(l_p \text{ or } L_n)}{(l_n \text{ or } L_p)}$$

Transition region

Contact potential $\psi = (kT/e) \ln [n_i^2/(N_A N_D)]$

$$\text{Width } w_t = \sqrt{\left[\frac{2\epsilon_0 \epsilon_r}{e} \left(\frac{1}{N_A} + \frac{1}{N_D} \right) (|\psi| - V_D) \right]}$$

Maximum field $E_{\max} = \dfrac{2|V_D + \psi|}{w_t}$

$$\text{Capacitance } C_t = A \sqrt{\left[\frac{e \epsilon_0 \epsilon_r}{2(1/N_A + 1/N_D)(|\psi| - V_D)} \right]}$$

Diffusion capacitance contributions

In the p-side, $C_{dp} = \dfrac{l_p^2 I_n}{2D_n} \dfrac{e}{kT} \quad$ or $\quad (t_n I_n) \dfrac{e}{kT}$

In the n-side, $C_{dn} = \dfrac{l_n^2 I_p}{2D_p} \dfrac{e}{kT} \quad$ or $\quad (t_p I_p) \dfrac{e}{kT}$

Diode slope resistance:

$$r_e = \frac{(kT/e)}{I_D}$$

MOSFETs

Threshold voltage

$$V_T = \frac{\left| \dfrac{Q_D}{A} \right| \pm \left| \dfrac{Q_{SS}}{A} \right|}{C_{ox}} + |\psi_D|$$

$$\psi_D = \frac{2kT}{e} \ln \left(\frac{N_{A/O}}{n_i} \right)$$

$$\frac{Q_D}{A} = \pm \sqrt{(2\epsilon_0 \epsilon_r e \psi_D N_{A/D})}$$

$$\frac{Q_{SS}}{A} = e(N_{ox} \pm N_{sd})$$

$$C_{ox} = 4\epsilon_0/t_{ox} \text{ for silicon oxide}$$

Characteristic equations

Linear region, $V_{DS} < V_{GS} - V_T$,

$$I_D = \beta V_{DS}[(V_{GS} - V_T) - \tfrac{1}{2}V_{DS}]$$

Saturated region, $V_{DS} > V_{GS} - V_T$,

$$I_D = \tfrac{1}{2}\beta(V_{GS} - V_T)^2[1 + \lambda(V_{DS} + V_T - V_{GS})]$$

$$\beta = W\mu C_{ox}/L \qquad (\mu \text{ surface value})$$

Small-signal parameters in saturated region

$$g_m \approx \sqrt{(2\beta I_D)}$$

$$g_{ds} \approx \frac{I_D}{V_{DS} + 1/\lambda}$$

BJTs

$$I_C + I_B = I_E$$

$$\beta = I_C/I_B \approx I_E/I_B$$

Small-signal parameters

$$r_i = \beta r_e$$

$$g_m \approx 1/r_e$$

$$g_o = \frac{I_C}{V_{CE} + V_A}$$

Optical devices

Photon energy $E_{ph} = hf = hc/\lambda_0$

Absorption $N_x = N_0 \exp(-\alpha x)$

Diode current $I_D = I_s[\exp(eV_D/kT) - 1] - Ip$

Responsivity $R = \dfrac{I_p}{p} = \dfrac{\eta P \lambda_0}{hc}$

Quantum efficiency $\eta_{int} = \dfrac{1}{1 + t_{rr}/t_{nr}}$

Material wavelength $\lambda_m = \lambda_0/n$

Laser cavity resonance $\dfrac{N\lambda_m}{2} = L$

Index

Terms that appear frequently in the text are listed for the page where they are introduced, explained or defined.